*Biological Effects
of Radiation*

Biological Effects
of Radiation

Second Edition

J. E. Coggle

Department of Radiobiology,
The Medical College of St Bartholomew's Hospital, London

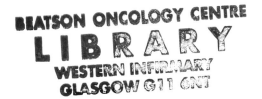

International Publications Service
Taylor & Francis Inc.
New York
1983

First edition published 1971 by Wykeham Publications (London) Ltd
Reprinted 1973 and 1977

This edition published 1983 by International Publications Service,
Taylor & Francis Inc., 114 East 32nd Street, New York, N.Y. 10016

Typeset by Georgia, Liverpool L3 9EG
Printed and bound in Great Britain by
Taylor & Francis (Printers) Ltd, Rankine Road,
Basingstoke, Hampshire RG24 0PR

ISBN 0-8002-3075-2

Library of Congress Catalog Card No: 83-80179

Contents

Preface

Since we are all exposed throughout our lives to ionizing radiation from natural and artificial sources, it is important for us to have an understanding of the interaction of such radiation and living matter. The term 'ionizing radiation' is taken to include X and γ rays, α and β particles, electrons, protons, neutrons and cosmic rays. This book is therefore not concerned with ultra-violet and visible light, infra-red radiation and radio waves since these produce no ionizations in living matter. Ionization is the process by which a fast-moving quantity of energy is transferred to some of the atoms of the material through which it is travelling, leaving them as electrically charged ions. The physico-chemical changes caused by the ionization of the atoms of living matter occur in a fraction of a second, whereas the processes whereby these physico-chemical changes eventually lead to such biological changes as genetic mutations, cell death and cancer may take hours, months, or even decades. The links between the physico-chemical and the biological effects of ionizing radiation are poorly understood and are the subject matter of basic scientific research in radiobiology, involving teams of physicists, chemists, biologists and medical scientists. Scientifically, the fascination of radiation biology lies in the attempt to explain how the small amounts of absorbed radiation energy can have such widespread biological results.

Besides the purely scientific reasons, there are practical motives for the study of the biological effects of ionizing radiation. They stem from the widespread and increasing use of radiation in modern society that makes it imperative for us to have a precise evaluation of the biological risks involved in its use. The following précis of some of the scientific, medical, agricultural, industrial and military uses of radiation illustrates this point.

In many branches of physics, chemistry and biology use is made of radiation from radio-isotopes. In agriculture and horticulture, radiation is often used to produce new varieties of crops and flowering plants; and a modern method of pest control involves the release of radiation-sterilized male animals that mate unsuccessfully with the females. The use of radiation from radio-isotopes in the diagnosis and treatment of disease is known as 'nuclear medicine'. X-ray machines are used to make internal organs visible and to diagnose disease. More powerful sources of radiation, high-energy X-ray machines, particle accelerators and cobalt-60 γ-ray sources are used in radiotherapy to treat disease.

In industry, radiation is used to detect microscopic cracks and cavities in metal structures, the presence of such flaws causes a detectable increase in the amount of radiation capable of traversing a structure. In the food industry, very high doses of radiation are being investigated as a method of sterilization; the radiation killing the bacteria, fungi and other micro-organisms in the food that might otherwise cause disease. A similar sterilization procedure is used routinely for medical instruments such as syringes.

Energy from the fission of uranium and plutonium is harnessed in nuclear power stations and is used to drive ships and submarines. The military use nuclear fission and fusion to produce nuclear weapons and hydrogen bombs. These weapons produce devastating immediate effects from radiation and from the shock-wave of the explosion, and they create the long-term hazard of radiation 'fall-out'. High-altitude flight and particularly space flight involves exposure to extra-terrestrial (cosmic) radiation. And finally, in the home, television, especially colour sets, infra-red cookers and luminous watches all produce minute amounts of ionizing radiation.

The sensible use of radiation in laboratories, hospitals or industry necessitates an awareness by personnel of its potentialities, its limits and its hazards. It is the task of the radiation biologists to study the immediate and long-term somatic and genetic hazards associated with radiation exposure and also to investigate the likely global effects on plants and animals of any increase in the natural (environmental) background radiation caused by fall-out from nuclear weapons tests.

It is impossible to cover the whole radiobiology in a book of this size and consequently most of the radiation effects on micro-organisms and plants are excluded.

The effects of radiation on mammals, and in particular man, have been most intensively explored, and these studies form the main contents of the book.

Preface to the Second Edition

There have been numerous advances in our knowledge of radiation biology in the 10 years or so since the first edition of this book. I have attempted to incorporate some of them into this edition, particularly in Chapters 2, 7 and 9. The increasing emphasis on radiation protection and the concerns over the risk of low levels of radiation is evident in the much expanded Chapters 9 and 11. Finally, I have added Chapter 12 on the topical issues surrounding nuclear power and the environment.

The units used throughout the book are SI units.

Chapter 1
Some properties of ionizing radiation

1.1. Introduction

This book is concerned with the interaction of ionizing radiation and living matter. The biological effects of ionizing radiation are caused by the absorption of the radiation energy in the tissues and by the distribution of that energy in the tissues. If radiation were to pass straight through living material without leaving any energy behind, it would have no biological effect. To view the radiations most frequently used in radiobiological studies in perspective, these are presented in tables 1.1 and 1.2, together with their most important properties. Table 1.1 refers to particulate ionizing radiation while table 1.2 refers to the range of ionizing electromagnetic radiation used.

The different effects observed for these very different radiations may be closely linked with their energy deposition characteristics. These, in turn, depend on the mass, charge and energy of the radiation considered.

This first chapter outlines some of the basic physical and chemical steps that precede the biological effects of radiation. It will not be a comprehensive treatment of these vast fields and it assumes a basic understanding of atomic structure and the fundamental properties of alpha (α), beta (β) and gamma (γ) rays.

1.2. Radioactivity and nuclear reactions

Matter is made up of elements, of which there are 92 naturally occurring ones plus a few that have been artificially produced. Elements are composed of atoms which consist of a positively charged nucleus surrounded by negatively charged orbiting electrons. The nucleus is composed of protons (positively charged) and neutrons (electrically

1

Table 1.1. Particulate radiation used in radiobiological studies

Radiation	Rest mass (kg)	Charge (C)	Approximate energy used (eV)	(J)
α particle			~1 MeV	$1 \cdot 6 \times 10^{-13}$
or	$6 \cdot 7 \times 10^{-24}$	$+3 \cdot 2 \times 10^{-19}$	$\updownarrow$	
He nucleus			~200 MeV	$3 \cdot 2 \times 10^{-12}$
β particle			~10 keV	$1 \cdot 6 \times 10^{-15}$
or	$9 \cdot 1 \times 10^{-31}$	$-1 \cdot 6 \times 10^{-19}$	$\updownarrow$	
electron e$^-$			~15 MeV	$2 \cdot 4 \times 10^{-12}$
Neutrons	$1 \cdot 7 \times 10^{-27}$	0	*slow*	
			0·025 eV	$4 \cdot 0 \times 10^{-21}$
			$\updownarrow$	
			0·1 keV	$1 \cdot 6 \times 10^{-17}$
			intermediate	
			0·10 keV	$1 \cdot 6 \times 10^{-17}$
			$\updownarrow$	
			0.02 MeV	$3 \cdot 2 \times 10^{-15}$
			fast	
			⟩0·02 MeV	⟩ $3 \cdot 2 \times 10^{-15}$
			~1 MeV	$1 \cdot 6 \times 10^{-13}$
Protons	$1 \cdot 7 \times 10^{-27}$	$+1 \cdot 6 \times 10^{-19}$	$\updownarrow$	
			~30 GeV	$4 \cdot 8 \times 10^{-9}$
			~1 MeV	$1 \cdot 6 \times 10^{-13}$
Deuterons	$3 \cdot 3 \times 10^{-27}$	$+1 \cdot 6 \times 10^{-19}$	$\updownarrow$	
			~200 MeV	3.2×10^{-11}
Heavy ions				
^{7} Li	$1 \cdot 2 \times 10^{-26}$	$+4 \cdot 8 \times 10^{-19}$		
^{11} B	$1 \cdot 8 \times 10^{-26}$	$+8.0 \times 10^{-19}$	~1 MeV	$1 \cdot 6 \times 10^{-13}$
^{12} C	$2 \cdot 0 \times 10^{-26}$	$+9 \cdot 6 \times 10^{-19}$	$\updownarrow$	
^{14} N	$2 \cdot 3 \times 10^{-26}$	$+1 \cdot 1 \times 10^{-18}$		
^{16} O	$2 \cdot 7 \times 10^{-26}$	$+1 \cdot 3 \times 10^{-18}$	~200 MeV	$3 \cdot 2 \times 10^{-11}$
^{20} Ne	$3 \cdot 3 \times 10^{-26}$	$+1 \cdot 6 \times 10^{-18}$		
^{40} Ar	$6 \cdot 7 \times 10^{-26}$	$+2 \cdot 7 \times 10^{-18}$		
$\pi+$(pions)		$+1 \cdot 6 \times 10^{-19}$		
	$2 \cdot 5 \times 10^{-29}$		~100 MeV	$1 \cdot 6 \times 10^{-11}$
$\pi-$(pions)		$-1 \cdot 6 \times 10^{-19}$		

neutral particles). Protons and neutrons have the same mass and are approximately 1800 times heavier than electrons. Atoms are normally electrically neutral, the number of electrons balancing the number of protons in the nucleus. The atomic number of an element, the number of protons in the nucleus (Z) and the mass number (A), are shown as the sub- and super-scripts to the left of the symbol for the element. Thus $^{131}_{53}$I is the atom of iodine-131 with 53 protons and 78 neutrons. Particular values of A and Z define particular nuclides and there are approximately 1600 known nuclides. The different forms of an element, i.e., atoms with the same atomic number but different mass numbers, are called isotopes, for example, iodine-131 and iodine-125. Most nuclides are stable, while

Table 1.2.   Electromagnetic radiation. X-rays and γ rays. Approximate energy ranges of interest

Photon energy (eV)	(J)	Frequency (Hz)	Wavelength (nm)	Properties
124 eV to	$2\cdot0\times10^{-17}$	$3\cdot0\times10^{16}$	10	Soft X-rays from excitations of inner electrons; very small penetration
$12\cdot4$ keV	$2\cdot0\times10^{-15}$	$3\cdot0\times10^{18}$	$0\cdot1$	
$12\cdot4$ keV to	$2\cdot0\times10^{-15}$	$3\cdot0\times10^{18}$	$0\cdot1$	Diagnostic X-rays and superficial therapy
124 keV	$2\cdot0\times10^{-14}$	$3\cdot0\times10^{19}$	$0\cdot01$	
124 keV to	$2\cdot0\times10^{-14}$	$3\cdot0\times10^{19}$	$0\cdot01$	Deep therapy X-rays, and γ rays from many radioactive isotopes, e.g. Co^{60}
$1\cdot24$ MeV	$2\cdot0\times10^{-13}$	$3\cdot0\times10^{20}$	$0\cdot001$	
$12\cdot4$ MeV	$2\cdot0\times10^{-12}$	$3\cdot0\times10^{21}$	$0\cdot0001$	Radiation from small betatron
124 MeV	$2\cdot0\times10^{-11}$	$3\cdot0\times10^{22}$	$0\cdot00001$	Radiation from large betatron
$1\cdot24$ GeV	$2\cdot0\times10^{-10}$	$3\cdot0\times10^{23}$	$0\cdot000001$	Radiation produced in large synchrotrons

Modified after H. E. Johns, 1964, *The Physics of Radiology,* 2nd edition, Charles C. Thomas; courtesy the author and publisher.

some are unstable and undergo radioactive decay to produce more stable configurations. Each radionuclide decays in a characteristic way producing one or more emissions of a defined energy. There are several types of decay, including α, β and γ emissions, internal conversion and electron capture.

When a nucleus emits an α particle its atomic number falls by 2 and its mass number by 4. For example, the decay of uranium-238 by α-emission produces an isotope of thorium:

$$^{238}_{92}U \quad \longrightarrow \quad ^{234}_{90}Th + ^{4}_{2}He$$

$$\text{uranium-238} \quad \longrightarrow \quad \text{thorium-234} + \alpha\,\text{particle}$$

the α particle (a helium nucleus) has an energy of about $4\cdot18$ MeV.†

Most α-emitters are elements of high atomic number and the energy of the α particles ranges from 4 to 9 MeV.

Beta particles, 'electrons', are emitted from the nucleus by the conversion of a neutron into a proton. Therefore, when an atom emits a β particle the atomic number rises by one and the mass number is un-

---

†The electron volt (eV) is a measure of energy equal to $1\cdot6\times10^{-19}$ J. 1 MeV is 10^6 eV.

changed. For example, iron-59 decays by β emission to produce cobalt-59, the stable isotope of cobalt:

$$^{59}_{26}\text{Fe} \rightarrow {}^{59}_{27}\text{Co} + {}^{0}_{-1}\text{e}$$

$$\text{iron-59} \rightarrow \text{cobalt-59} + \beta \text{ particle}$$

Note that the β particle has atomic and mass numbers of -1 and 0, respectively. It has a zero mass number because it contains no protons or neutrons and an atomic number of -1 as it has a single negative charge. After the emission of an α or β particle the nucleus may be in an excited state and will rearrange itself by immediately releasing this excitation energy in the form of γ rays. Such decay produces no change in either atomic or mass number as γ rays are shortwave electromagnetic radiation. For example, in the decay of cobalt-60 by β emission, two γ rays of $1 \cdot 332$ MeV and $1 \cdot 172$ MeV are produced:

$$^{60}_{27}\text{Co} \rightarrow {}^{60}_{28}\text{Ni} + {}^{0}_{-1}\text{e} + \gamma$$

$$\text{cobalt-60} \rightarrow \text{nickel-60} + \beta + \gamma \text{ rays}$$

Alternatively, the transition from the excited state may be achieved by a process called 'internal conversion' in which the γ ray's energy is transferred to one of the inner orbital electrons. The electron is ejected with an energy equal to that of the γ ray minus the electron's binding energy. After internal conversion, characteristic X-rays will be emitted as the orbital electrons rearrange themselves to fill the vacancy left by the converted electrons. If any of the X-rays themselves are absorbed by the orbital electrons there may be further emission of electrons called 'Auger electrons'.

A further type of decay-electron capture occurs in nuclides when the ratio of neutrons to protons is too low for nuclear stability. A proton in the nucleus captures an electron from one of the inner orbits – usually the innermost, 'K shell'. The proton is thereby converted to a neutron and a neutrino is emitted. Once again, characteristic X-rays are always emitted as the vacancy in the K shell is filled by a rearrangement of the orbital electrons.

The decay of a radioactive sample is statistical, but follows a strict pattern. The number of disintegrations per second is proportional to the activity present at any instant.

$$\frac{dA}{dt} = -\lambda A$$

which integrating with respect to time gives

$$A = A_o \exp(-\lambda t)$$

where A_o=activity at time 0; A=activity at time t; and λ is the disintegration constant.

This exponential decay of activity has an associated 'half life' which is defined as the time taken for the activity to decrease to one half of its value at any given time.

The half life $(t_{\frac{1}{2}})$ may be found substituting $\frac{1}{2}A_o$ for A in the last equation

$$\tfrac{1}{2}A_o = A_o \exp\left(-\lambda t_{\frac{1}{2}}\right)$$

which taking logs gives $t_{\frac{1}{2}} = (\log_e 2)/\lambda$

The half life of radio-isotopes can range from less than a micro-second to greater than 10^{11} years, which is greater than the age of the earth!

1.3. Nuclear fission

Until 1934 it was thought that uranium with 92 protons was the element with the highest atomic number, but it was then found that if uranium was bombarded with neutrons it would pick up a neutron and emit a β particle, so becoming element 93, neptunium. This process of neutron activation can be used to obtain elements of even higher atomic number – plutonium (94), americium (95), curium (96), berkelium (97), californium (98), and so on. These elements are called the transuranic elements or higher actinides.

Some of these heavy nuclides, for example, uranium-235, curium-242 and californium-252, are not only radioactive and emit particles but they may also undergo 'fission'. In this type of decay, the nucleus spontaneously divides into two approximately equal parts. At the same time large amounts of energy are emitted, mostly as the kinetic energy of the fission fragments, but some is carried off by neutrons and γ rays that accompany the fission. The new nuclides ('daughters') produced by a fission reaction range from about barium to bromine in the periodic table and they are invariably unstable and decay by β emission. The fission yield varies with mass number in what is often called a saddleback distribution. Figure 1.1. shows this characteristic distribution yield and some of the most important fission products from the disintegration of ^{235}U. An average of 2·5 neutrons (between 1 and 5 neutrons) are emitted during this fission process. For example, one possible fission of ^{235}U involves the release of four neutrons:

$$^{235}_{92}\text{U} + \text{n} \longrightarrow {}^{135}_{52}\text{Te} + {}^{97}_{40}\text{Zr} + 4\text{n}$$

These neutrons can produce further fissions producing a self-sustaining chain reaction. In a nuclear reactor, the fissions are induced by neutrons rather than by relying on the spontaneous fission of such unstable nuclides as uranium, curium or californium. Over 99 per cent of the neutrons emitted in a chain reaction are released within 10^{-14} seconds (prompt neutrons); the rest (~0.75%) are emitted over a period of several minutes (delayed neutrons). The latter are very important in allowing the

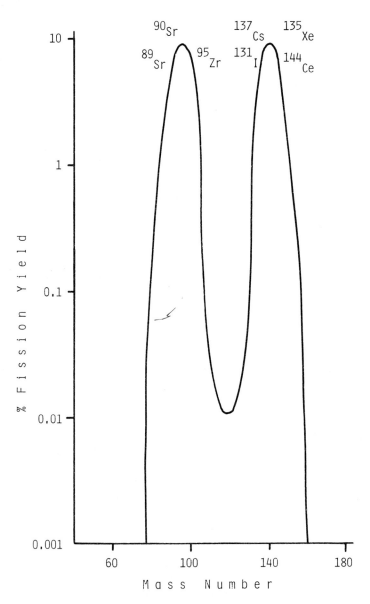

Figure 1.1.   **Characteristic distribution yield of fission products of ^{235}U**

control of the fission reaction that is essential in the nuclear power industry (see Chapter 12). An uncontrolled chain reaction that is contained is the basis of the atom bomb.

1.4. The physical changes that follow the absorption of ionizing radiation

The absorption of alpha particles

The α particle is composed of two neutrons and two protons. It is therefore relatively massive, being some 7500 times as heavy as an electron. On passing through matter, such a positively charged particle exerts a strong attractive force on the negatively charged orbital electrons of the atoms near its path. This attraction may pull off one or more of these electrons, and the energy required for such an interaction causes the dissipation of some of the energy of the α particle. The electron thus removed and the positive ion left behind constitute an 'ion pair' and the process is called 'ionization'. The interaction between the α particle and the atoms of the media through which it is passing may not always be strong enough to cause ionization, but may cause 'excitation'. Excitation differs from ionization in that it does not involve the removal of electrons from the atoms of the media, but is merely the raising of some of the orbital electrons to higher energy levels within the atom. Excitations and ionizations are the most important ways in which α particles (or any ionizing radiations) transfer their energy to the matter through which they may be passing.

It has been shown experimentally that in air it requires an average of about 34 eV to form an ion pair; this energy is supplied from the ionizing particle's kinetic energy. A considerable fraction of the 34 eV is dissipated as excitation energy.

Ionization events can be demonstrated in a cloud chamber or on a photographic plate. These methods show that the ionization track of the α particle is dense and straight. The number of ion pairs formed per unit length of the α particle track is called its 'linear ion density' or 'specific ionization', and it depends on the energy and charge of the particle and on the density (and atomic number, Z, of the elements) of the material through which it is travelling. First, the number of ion pairs produced is directly proportional to the initial energy of the particle. Second, the rate at which the energy is lost along the track of a particle is proportional to the square of the charge of the particle. An α particle (2 positive charges) will lose energy four times as fast as a proton (1 positive charge), provided that both have the same velocity. Third, the velocity of a particle also governs the rate at which it loses energy, since it determines the time interval for which a particle is able to exert its electric field on the atoms of the media. A faster moving particle is less likely to ionize a substance through which it is passing than a slower one. As a result of their mass, α particles move relatively slowly and so have ample time to ionize the atoms of the media through which they pass. Because of their charge (2+) and their slowness they give up all their energy in short, dense, straight tracks which have a specific ionization pattern, as shown in figure 1.2. As the α particle penetrates deeper into the media more and more interactions (ionizations and excitations) occur, thus reducing its speed, which in turn increases the chances of further interactions. Eventually, a peak, often referred to as the Bragg ionization peak, is reached, followed by a decline to zero, when all the α particle's energy has been dissipated. The exhausted particle then attracts two electrons to itself and becomes a neutral helium atom. The range of the 1 MeV α particle is a few centimetres in air at atmospheric pressure. In living tissue, because of its greater density, the 1 MeV α particle is only able to travel a few tens of micrometres. Since α particles deposit their energy in such short tracks they can be very damaging biologically if they get inside the cell, as may be the case for α-emitting radio-isotopes.

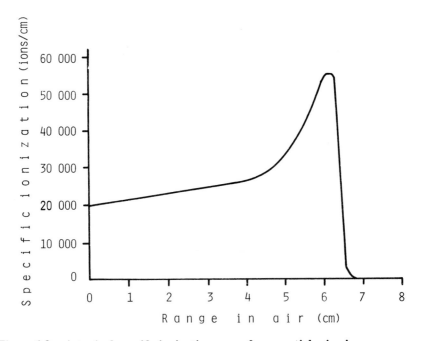

Figure 1.2.   A typical specific ionization curve for α particles in air
Source: C. H. Wang and D. L. Willis, 1965, *Radiotracer Methodology in Biological Sciences*, Prentice-Hall Inc.; courtesy the publisher.

The absorption of electrons

The interactions between electrons (or β particles as they are called if they are produced in radioactive decay processes) and matter involve exactly the same processes of excitation and ionization as described for the α particle, and result in the gradual loss of energy of the impinging electron. Due to its small mass and to its single negative charge, each time an electron approaches close to orbital electrons it is deflected from its path. It will also be deflected from its path by the positive atomic nuclei. As a result of these processes the path of the electron is tortuous and difficult to define, in contrast to the straight track of the α particle. Because of the tortuous path, the depth to which the electrons penetrate (their range) will be less than their true path length (see figure 1.3).

Figure 1.3. **Illustration of the difference between the depth of penetration of an electron and its true path length**

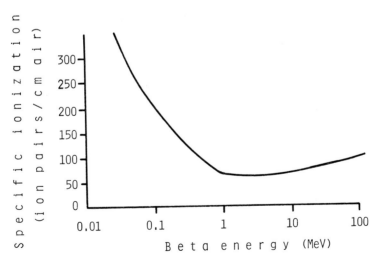

Figure 1.4. **The relationship between the specific inonization and the energy of electrons**
Source: C. H. Wang and D. L. Willis, 1965, *Radiotracer Methodology in Biological Sciences,* Prentice-Hall Inc.; courtesy the publisher.

Biological Effects of Radiation

Because of the devious path of an electron in a beam it is not possible to plot a curve of the specific ionization along the track of an electron as it is for the α particle. However, a curve of the relationship between the energy of an electron and its specific ionization can be plotted, as shown in figure 1.4. This graph (reading from right to left) shows that the greatest density of ionizations will occur at the end of the track of an electron, when its energy has been degraded to a few hundred electron volts. As the energy of the electron falls so does its velocity, which increases the probability of the interactions between it and the atoms of the material through which it is passing.

The absorption of neutrons

Neutrons carry no electric charge and therefore their interactions with matter result from direct collision processes with atomic nuclei. These interactions are rare events and dependent on the energy of the neutrons, the atomic density, and the masses of the atoms involved; as a consequence neutrons can penetrate deep into matter.

Slow or thermal neutrons ($0 \cdot 025$ eV to $0 \cdot 1$ keV) interact mainly by entering the atomic nuclei and being held or 'captured' there. Fast neutrons ($>0 \cdot 02$ MeV) interact mainly by elastic collisions with the nuclei. Maximum energy transfer occurs for a head-on collision, and simple mechanics indicates that if the masses of the two particles involved in the

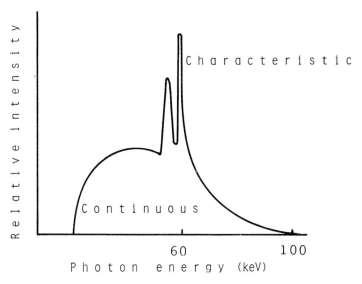

Figure 1.5. Continuous X-ray spectrum with characteristic lines for a tungsten target bombarded with 100 keV electrons

collision are equal, then total transfer of energy is possible. For the neutron, this condition is satisfied when it collides with a hydrogen nucleus (i.e. a proton); and in living tissue with its high density of hydrogen atoms this interaction is of great importance. Recoil protons of energies up to that of the incident neutron are produced, and being heavy charged particles, they cause intense ionization as they are slowed down (cf. α particles). Neutrons also collide with other atomic nuclei (e.g. carbon, oxygen, etc.), producing heavily ionizing radiation resulting in considerable biological damage.

Neutrons of intermediate energy (0·1 – 20 keV) interact through both capture and collision processes.

The absorption of photons

X-rays and γ rays are electromagnetic radiations consisting of streams of energetic photons (packets of energy) with the capacity to produce ionizations. In order to get a clear picture of their interaction with matter it is useful first to examine the way in which X-rays are produced.

X-rays arise when fast electrons are arrested by a target. The intensity of the X-rays increases with the atomic number (Z) of the target, so that it would seem logical to use uranium with an atomic number of 92. However, a material with a high melting point is required, so a compromise such as tungsten ($Z = 74$) or gold ($Z = 79$) is used. The X-ray spectrum produced for a tungsten target is indicated in figure 1.5. It consists of two parts: a continuous background radiation and, superimposed on this, peaks—the characteristic radiation of the metal. The continuous Bremsstrahlung (from the German 'braking') or 'white' radiation arises when an incident electron undergoes a Coulomb interaction scattering due to close collision with a target nucleus. In these types of collision considerable accelerations occur and the electron may be scattered through large angles. The electron loses energy by electromagnetic radiation, the energy of which depends on the electron–nucleus interaction. Photons with a range of energies are obtained, so that we get a continuous range of wavelengths—a continuous spectrum (see figure 1.5).

The characteristic radiation may be explained by considering the interaction of the incident electron with an atom, based on the Bohr model. The model of an atom with nucleus of positive charge Z units and the same number Z of associated electrons, postulates that the latter move in certain allowed orbits around the nucleus, and while in these orbits they do not radiate electromagnetic radiation. If, however, an electron moves from an orbit where it has energy E_i to an orbit where it has lower energy E_f then electromagnetic radiation is emitted whose energy $h\nu$ is given by

$$h\nu = E_i - E_f$$

where h is Planck's constant and ν is the frequency of the radiation.

Further, the number of electrons which can populate a given orbit is governed by quantum mechanical rules and the Pauli Exclusion Principle, which states that no two electrons can be in the *same* quantum state in any one system.

The orbit closest to the nucleus which has the greatest binding energy is known as the K shell, the next the L shell and so on. The binding energies (E_b) for heavy elements are much higher than those for lighter elements such as carbon and oxygen (e.g., E_b for the K shell of tungsten $= 70$ keV, while E_b for the K shell of carbon $\approx 0\cdot3$ keV). This fact has an important bearing on the interaction of photons with matter (see The photoelectric effect, below; Compton scattering, p. 14).

The incident electron may collide with an electron in one of the atomic shells of the atom causing it to be ejected. If E_b is the binding energy, E_1 the energy of the incident electron and E_2 the energy of the ejected electron, then the final energy of the incident electron after interaction, E_3, is given by

$$E_3 = E_1 - (E_2 + E_b)$$

When the hole in the atomic shell is filled either by an electron from an outer shell or by an external electron, electromagnetic radiation characteristic of the transition taking place is emitted. This results in the characteristic lines (often referred to as the K, L, M, etc. lines) observed in the X-ray spectrum (see figure 1.5).

Gamma rays are similar in nature to X-rays but occur during the transition of a nucleus from an excited state to a more stable one. Since γ rays are emitted during allowed, discrete transitions, γ ray spectra are line spectra, with no continuous background, as is found for X-rays. Apart from this difference their interactions with matter are similar to those of X-rays.

The interactions of X-rays and γ rays with matter may be divided into three main categories, whose relative importance depends on the energy of the photons and the atomic number of the matter. They are known as the photoelectric effect, Compton scattering and pair production. The first two involve interactions of the photons with the electrons of the traversed material, while the last is an absorption event occurring within the strong nuclear field of the atom. There are other interactions, but these are quantitatively of little importance in a radiobiological context.

The photoelectric effect

This is the predominant mode of absorption in tissue for photons of low energy ($\langle 25$ keV). This interaction is similar to that already discussed in the production of characteristic X-rays by electrons. Here the *photon* interacts with a bound electron and is absorbed, causing the ejection of the electron from its shell with energy given by

$$E_e = h\nu - E_b$$

where $h\nu$ = photon energy; E_b = binding energy; E_e = electron energy.

After a short time another electron fills the vacated place with the emission of characteristic radiation. Figure 1.6 illustrates the photoelectric effect in tungsten irradiated by 75 keV X-rays, where an electron is removed from the K shell ($E_b = 69 \cdot 5$ keV) and electrons 'jump' from the *L, M, N* shells consecutively to fill the shell holes and characteristic radiation is emitted; note the high-energy photon in the transition from *L* to *K* ($58 \cdot 5$ keV). The situation is quite different in the element carbon (prominent in tissue). Here the K shell has a binding energy of about $0 \cdot 3$ keV, so that for an incident 75 keV photon, $74 \cdot 7$ keV is given to the emitted K electron, while the photon ultimately emitted in filling the K shell has an energy of only $0 \cdot 3$ keV. This is immediately absorbed. Thus most of the energy goes to the photoelectron which in turn produces ionizations and excitations.

The photoelectric effect for low-energy X-rays ($\langle 0 \cdot 5$ MeV) represents an

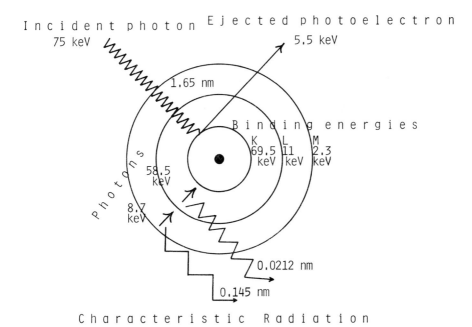

Figure 1.6. The photoelectric effect in tungsten

interaction with electrons bound in the atomic network; the photon is completely absorbed and an electron is emitted. Subsequently characteristic radiation is produced.

Compton scattering

As the energy of the incident photon beam increases from 25 keV to about 25 MeV the most important type of interaction in tissue producing attenuation and absorption is Compton scattering. Here the photon interacts with the outer electrons of the atom, which have low binding energy and may be considered effectively as free electrons. This is particularly true of the light elements of soft tissue where the binding energy is very low (<1 keV). As the photon energy approaches the megavolt range even the K shell electrons of heavy elements, for example lead ($E_b = 88$ keV), may be considered as free electrons.

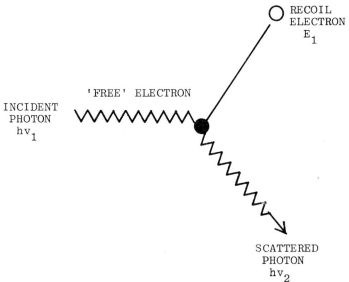

Figure 1.7. Compton scattering

The interaction is an inelastic collision, where the free electrons and photons may be considered as shown in figure 1.7.

No energy is lost in the collision so that energy E_1 is given to the 'recoil' electron and the photon leaves with diminished energy $h\nu_2$

$$h\nu_1 = h\nu_2 + E_1$$

The scattered photon may interact further with matter either by Compton scattering or the photoelectric effect depending on its residual energy.

Pair production

The rest mass of an electron is $0 \cdot 51$ MeV and at photon energies in excess of $1 \cdot 02$ MeV, pair production occurs, a striking example of Einstein's principle of the equivalence of mass and energy ($E = mc^2$).

In the strong electric field surrounding the nucleus the photon disappears, being converted into an electron and a positron (which has the same mass as an electron but a positive charge), thus conserving charge.

Any photon energy in excess of $1 \cdot 02$ MeV is shared between the particles as kinetic energy (see figure 1.8). Thus,

$$h\nu = 1 \cdot 02 \, \text{MeV} + E_{ELECTRON} + E_{\text{POSITRON}}$$

This seemingly extraordinary effect obeys the laws of conservation of charge, energy, mass and momentum as well as the more sophisticated laws of quantum mechanics. The positron, surrounded by a sea of electrons as it traverses the absorber, is in considerable danger of annihilation, and after it has been slowed down by interactions similar to those of energetic electrons it collides with an electron of the absorber and is annihilated.

This annihilation represents the reverse of the initial energy to mass conversion, and two photons of energy $0 \cdot 51$ MeV, known as 'annihilation radiation' are emitted at 180° to one another (figure 1.8). Thus charge, momentum and energy are again conserved.

The energy of the incident photons will determine which of the three processes, photoelectric effect, Compton scattering or pair production, plays the dominant role in their absorption. It is seen that ionizing electrons are produced in the interactions of X-rays and γ rays with matter—without these interactions the photons would travel straight through matter at the speed of light. This means that X- and γ rays have no absolute penetration depth or range. The depth of penetration of a

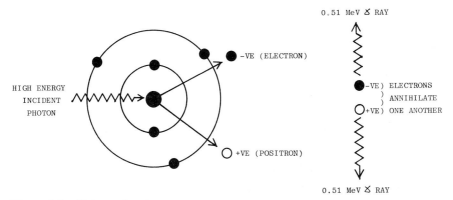

Figure 1.8.   Pair production

photon in a material depends on the energy of the photon, the density of the material and the atomic number (Z) of the atoms of the material. What happens is that as a result of the complex, random, interactions with matter the beam of photons gets progressively weaker in its intensity as it passes deeper and deeper into the matter. The beam becomes attenuated.

The absorption of X- and γ rays may be described by Beer's law

$$\Delta I = -\mu I \Delta x$$

where ΔI = change in beam intensity; Δx = thickness of material traversed; μ = linear absorption coefficient. Integrating this equation

$$I = I_0 e^{-\mu x}$$

where I_0 is the initial beam intensity; I is the beam intensity at thickness x; e is the base of natural logarithms (2·718).

The parameter μ is the result of the three interactions described above: photoelectric effect, Compton scattering and pair production.

Figure 1.9 shows a simple plot of the intensity of X- or γ rays against the thickness of the absorber. The logarithmic nature of the absorption curve is demonstrated in figure 1.10, which is a plot of the logarithm of the intensity versus the thickness of an absorber. The exponential curve is transformed into a straight line. The straight line presentation allows one to obtain accurately a quantity known as the 'half value thickness' or 'half value layer' (HVL), which is that thickness of absorber that will

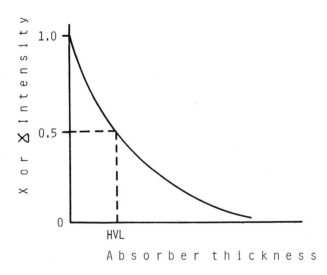

Figure 1.9. How the intensity of an X-ray or γ ray beam is reduced with increasing thickness of absorber

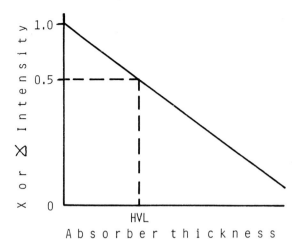

Figure 1.10. A semi-logarithmic plot of how the intensity of an X-ray or γ ray beam is reduced with increasing thickness of absorber

reduce the intensity of a beam of X- or γ rays by 50 per cent. The HVL is useful in estimating the thickness of shielding material that will give acceptable protection against X- or γ rays. For example, the HVL for X-rays from a 250 kV medical radiotherapy X-ray machine is about 3 mm of copper. So 6 mm of copper will reduce the intensity of the beam to 25 per cent (the first 3 mm reducing it to 50 per cent and the second 3 mm reducing the 50 per cent beam by 50 per cent, i.e. a total reduction of the initial beam strength by 75 per cent) leaving 25 per cent transmission.

1.5. The chemical changes that follow the absorption of ionizing radiation

We have so far considered the physical processes of the interaction of photons and ionizing particles with matter. These processes occur both in living and in non-living matter and they are over in the extremely short time of $10^{-24} - 10^{-14}$ seconds (see table 1.3). We must now examine the next step in the chain of events which may lead to a biological effect. Here an extrapolation must be made from the physical system previously discussed, where matter was considered as a sea of atoms, to the biological state where matter consists of molecules of various sizes, each composed of many atoms. The interaction of radiation with these molecules is assumed to be similar in nature to that with individual atoms. From henceforth the 'unit' considered will be the biological molecule. The chain of events in table 1.3 summarizes what occurs on the absorption of radiation by living matter and the approximate time-scale of the events.

Table 1.3.   Chain of events leading to radiation injury

Event	Timescale
1. *Initial interactions*	
Indirectly ionizing radiation [a]	$10^{-24} - 10^{-14}$s
Directly ionizing radiation [b]	$10^{-16} - 10^{-14}$s
2. *Physico-chemical stage*	
Energy deposition as primary track structure ionizations	$10^{-12} - 10^{-8}$s
3. *Chemical damage*	
Free radicals, excited molecules to thermal equilibrium	10^{-7}s – hours
4. *Biomolecular damage*	
Proteins, nucleic acids, etc.	ms – hours
5. *Early biological effects*	
Cell death, animal death	hours – weeks
6. *Late biological effects*	
Cancer induction, genetic effects	years – centuries

[a] X-rays, γ rays, neutrons
[b] Electrons, protons, α particles

So far we have dealt with the initial interactions of table 1.3 and shall mention a little about the physio-chemical stage later in this Chapter. The effect on biological molecules will be dealt with in Chapter 2 and early and late biological effects are the subject of the rest of the book.

A great deal is known about both the physical absorption of energy (stages 1 and 2) and about the biological effects of such absorption (stages 5 and 6). Less is known about the chemical and biomolecular events and still less is known about the connections between stages 2, 3, 4 and 5 and 6. It is not possible to fully describe the sequence of events that link the physical absorption of energy with the resulting biological effects. Nevertheless, we must examine some of the processes that occur at stage 3—those of free radicals and excited molecules.

Free radicals are electrically neutral atoms or molecules having an unpaired electron in their outer orbits. The concept of unpaired electrons can be explained using the Bohr model of the atom with its *K, L, M, N*, etc. shells, whose population is limited by the Pauli Exclusion Principle. The electron quantum state is determined by a number of parameters, one of these being spin, which may have two values only. In stable atoms and molecules, orbital electrons occur in pairs with opposite spin values. A free radical is formed by radiation when an atom is left with one of its outer orbital electrons unpaired with respect to spin. Free radicals are usually very reactive since they have a great tendency to pair the odd electron with a similar one in another radical or to eliminate the odd electron by an electron transfer reaction. Free radicals can therefore be electron acceptors (oxidizing species) or electron donors (reducing species).

Radiation produces excitations and ionizations at random, so that in a complex system such as living matter, those molecules that are most abundant are those most likely to become ionized. It follows that when living material, which is 70–90 per cent water, is irradiated, most of the absorbed energy will be taken up by the water molecules. Thus for an understanding of radiobiological effects the radiation chemistry of water is of extreme importance.

When pure water is irradiated it is ionized producing a fast moving free electron and a positively charged water molecule:

$$H_2O \xrightarrow{\text{radiation}} H_2\overset{+}{O} + e^-$$

This electron (e^-) will travel through the water, losing energy by various processes, until it is captured by another water molecule converting the latter into a negatively charged molecule:

$$e^- + H_2O \rightarrow H_2O^-$$

This process is a relatively slow one and the electron may become hydrated, i.e., surrounded by water molecules in such a way that the dipoles of several water molecules are orientated towards the negative charge of the electron. This assembly, the hydrated electron (e^-_{aq}), is stable enough at room temperature to give rise to a broad absorption spectrum with a maximum around 720 nm. Dry electrons are capable of reacting with a wide range of solute molecules. Such reactions are more likely at high solute concentrations ($\sim 0\cdot1-1\cdot0$ M) than in more dilute solutions where hydration of the electron is a competing process that is essentially complete in 10^{-11} s. Solute concentrations inside cells can be high enough so that dry electrons may be important.

Neither H_2O^- nor H_2O^+ is stable and each dissociates to give an ion and a free radical:

$$H_2O^+ \rightarrow H^+ + OH^•$$
$$H_2O^- \rightarrow H^• + OH^-$$

where the dot indicates the unpaired electron of the free radical.

For every 10^{-5} J of low linear energy transfer (LET) radiation energy absorbed by pure water the following new species are formed: 2·6 hydrated electrons, e^-_{aq}; 2·6 hydroxyl radicals, $OH^•$; 0·4 hydrogen atoms, $H^•$; and a small amount of H_2 and H_2O_2. The first three species being radicals are highly reactive and have lifetimes in the absence of other reactants or scavengers of up to several hundred microseconds. The absorption spectra of e_{aq}^- and $OH^\bullet$ are known.

The three mentioned species can react with each other or dimerize, and three such radical–radical interactions are given below:

$$H^\bullet + H^\bullet \rightarrow H_2$$
$$OH^\bullet + OH^\bullet \rightarrow H_2O_2$$
$$H^\bullet + OH^\bullet \rightarrow H_2O$$

They may react with other water molecules, for example

$$H_2O + H^\bullet \rightarrow H_2 + OH^\bullet$$

or the radicals may react with their own reaction products, for example

$$H_2O_2 + OH^\bullet \rightarrow H_2O + HO_2^\bullet$$

($HO_2^\bullet$ is the hydroperoxy radical)

The reactivities and rate constants for reactions of these species with many molecules have now been measured, mainly by pulse radiolysis techniques. Their chemistry is that of free radicals and they can therefore abstract hydrogen from organic molecules, RH,

$$RH + OH^\bullet \rightarrow R^\bullet + H_2O$$
$$RH + H^\bullet \rightarrow 0R^\bullet + H_2$$

These reactions result in new radical species. The primary as well as the secondary free radicals $R^\bullet$ can react with biologically important molecules and cause radiobiological damage. These reactions are generally held to be important in what is called the 'indirect' action of radiation. Indirect action involves aqueous free radicals as intermediaries in the transfer of radiation energy to biological molecules. In contrast, the direct action of radiation involves the simple interaction between the ionizing radiation and critical biological molecules. The latter become directly changed into free radicals as follows:

$$RH \xrightarrow{\hspace{0.5cm}\wedge\hspace{-0.2cm}\wedge\hspace{-0.2cm}\wedge\hspace{0.3cm}} RH^+ + e^- \rightarrow R^\bullet + H^+$$

Figure 1.11 illustrates the difference between the direct and indirect action of ionizing radiation.

From the point of view of biological damage it does not matter at all whether the critical molecule is damaged directly or indirectly. However, it does seem likely that much radiobiological damage is a consequence of indirect action, since cells and tissues are composed of approximately 70–90 per cent water. The use in figure 1.11 of the term 'target' molecule will be discussed in more detail in Chapter 3.

Free radicals may react with molecules of oxygen and such reactions are of great radiobiological significance because they may lead to the production of peroxide radicals both of hydrogen and of important organic

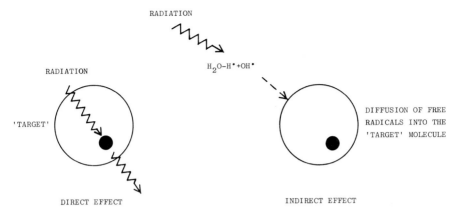

Figure 1.11.   Direct and indirect actions of ionizing radiation

molecules, some of which have been shown to be biologically damaging. The increased effectiveness of radiation in the presence of oxygen is known as the 'oxygen effect'. The increased yield of damaging free radicals formed in the presence of oxygen has been proposed to account for the effect. The reaction of oxygen with aqueous free radicals such as H and e^-_{aq} lead to the production of the relatively stable hydroperoxy radicals (HO_2) and to hydrogen peroxide:

$$O_2 + H^\bullet \quad \rightarrow \quad HO_2^\bullet$$
$$O_2 + e^-_{aq} \quad \rightarrow \quad O^-_2$$
$$O_2^- + H^+ \quad \rightarrow \quad HO_2^\bullet$$
$$2HO_2^\bullet \quad \rightarrow \quad H_2O_2 + O_2$$

Alternatively, if an organic biological molecule (RH) becomes a free radical either directly or indirectly it may interact with oxygen as follows:

$$R^\bullet + O_2 \quad \rightarrow \quad RO_2^\bullet \text{ (an organic peroxy radical)}$$

It can be seen that a chain reaction may be generated involving more RH:

$$RO_2^\bullet + RH \rightarrow RO_2H + R^\bullet$$

These reactions are tantamount to fixation of biological damage and occur at a rate thirty times faster than the competing reaction, e.g., $R^\bullet$ + cysteine or other hydrogen donor give RH, i.e. reconstitution. The important role of oxygen in radiobiology will be discussed in more detail in Chapter 8.

It is not necessary to go any further into the complex interactions of aqueous and other free radicals that are the basis of radiation chemistry. Many of these interactions lead to biologically harmful products and others lead to the initiation of damaging chain reactions, and many such reactions are enhanced by the presence of oxygen. For example, there is strong evidence that the OH• radical is involved in producing DNA single strand breaks, chromosome aberration, bacterial and mammalian cell killing.

1.6. The biological changes that follow the absorption of ionizing radiation

We have seen that, in order to affect matter, either living or non-living radiation must interact with it and cause excitations or ionizations of the atoms of the material. No biological damage will be caused by radiation that passes through a cell without depositing any of its energy. the dissipation of the energy of particulate radiation (electrons, protons, neutrons, α particles, etc.) is predominantly by direct ion pair production. The ionizations produced by X-rays, γ rays and neutrons are indirect in that they involve secondary electrons (photo- and recoil electrons) or recoil protons which in their turn ionize the material as they dissipate their energy. Ionizations are seldom produced singly but occur as double and triple events known as 'clusters' or 'spurs'. It is these ionization events which are held to be the principal cause of radiation effects in living matter—the exact role played by excitation in the production of biological damage is less well understood.

The enhanced effectiveness of radiation in the presence of oxygen (Chapter 8) is believed to be due to oxygen acting at the level of the initial chemical lesion, and a number of hypotheses have been put forward to explain this effect. One of these suggests that oxygen acts directly on the irradiated 'target' molecule and prevents the occurrence of any restoration processes. A more generally accepted hypothesis is that the

Table 1.4. Some of the types of mammalian radiobiological damage

Level of biological organization	Important radiation effects
Molecular	Damage to macromolecules such as enzymes, RNA and DNA, and interference with metabolic pathways
Subcellular	Damage to cell membranes, nucleus, chromosomes, mitochondria and lysosomes
Cellular	Inhibition of cell division; cell death; transformation to a malignant state
Tissue; organ	Disruption of such systems as the central nervous system, the bone marrow and intestinal tract may lead to the death of animals; induction of cancer
Whole animal	Death; 'radiation lifeshortening'
Populations of animals	Changes in genetic characteristics due to gene and chromosomal mutations in individual members of the species

presence of oxygen causes the appearance of more damaging free radical species and it is these that are held responsible for the oxygen effect.

The direct and indirect effect of radiation upon important biological molecules results in the wide range of biological effects seen in irradiated living organisms. The range and the complexity of the biological effects are treated in Chapters 2 and 12, although a selection of the types of radiobiological damage is given in table 1.4, which can be used for reference.

The quantity and the quality of the biological damage depend upon the dose of radiation, on the rate at which it is given, and on the distribution of the dose in the tissues. These three parameters will now be discussed.

1.7. The dose, the dose rate and the dose distribution of ionizing radiation

The dose

Until 1975, the two units of dosimetry in common use in radiation biology were the röntgen (R) and the rad.

The röntgen was the unit of exposure, i.e. the amount of radiation energy directed at a material and not that fraction actually absorbed. The röntgen is defined as the quantity of X- or γ radiation such that the associated secondary electrons emitted produce ions of either sign carrying a charge of $2 \cdot 58 \times 10^{-4}$ couloumb per kilogram of air. The associated electrons are the photoelectrons or recoil electrons. The röntgen was restricted to X- and γ radiation below 3 MeV, because of the difficulty of measuring the ionizations in air of the very energetic secondary electrons. It was because of these restrictions that the rad was adopted by the International Commission on Radiological Units in 1956. The rad (from *R*adiation *A*bsorbed *D*ose) was the most useful unit for radiobiological purposes since it was a measure of the radiation energy actually absorbed by the tissues. One rad was defined as the absorption of 10^{-2} joule of radiation energy per kilogram of material $(0 \cdot 01 \text{ J kg}^{-1})$.

Since the energy absorbed in tissue corresponding to an exposure to one röntgen is $0 \cdot 0095$ joules per kilogram, it follows that in tissue one röntgen gives an absorbed dose of $0 \cdot 95$ rad. So in practice these quantities were often considered interchangeable.

In the early 1970s, national and international discussions took place to introduce the Integrated System of Units (SI) into medical physics. In May 1975 the General Conference on Weights and Measures adopted new units. The röntgen is being phased out and there is no special unit for exposure in SI. The SI unit of absorbed dose is the gray (Gy) and it is defined as 1 joule per kilogram (1 J kg^{-1}) and is thus a hundred times larger than the rad (1 Gy = 1 J kg^{-1} = 100 rad). In Chapters 11 and 12 in discussing the problems of radiation protection and doses to radiation workers, the unit of dose equivalence, the sievert (Sv), has to be used. Unfortunately in some instances it is extremely difficult to convert grays to sieverts and so some passages contain both systems of units.

The unit of activity of a radioactive material was until recently the curie (Ci). The curie was based on the number of disintegrations occurring in 1 gram of radium in equilibrium with its products, and a radioactive material was said to have an activity of one curie if it exhibited $3 \cdot 7 \times 10^{10}$ disintegrations per second. The curie has been replaced by the SI unit the becquerel (Bq), which is defined as one disintegration per second. So 1 Ci=$3 \cdot 7 \times 10^{10}$ Bq. Since the becquerel is so small it necessitates the use of large multiple SI prefixes, for example, 1 megabecquerel (1 MBq)=10^6 Bq, 1 gigabecquerel (1 GBq)=10^9 Bq.

The measurement of radiation dose

Although we are primarily concerned with the energy absorbed by biological material, measured in grays, in practice this is very difficult. Consequently radiation doses are most often based on ionization effects in air using ionization chambers of varying types. Radiation produces ionizations in air and other gases and the ionization current can be measured by applying a potential difference between the two electrodes in a gas-filled chamber. The resulting electric current between the two electrodes is a measure of the amount of ionization the radiation has produced in a defined volume of the ionization chamber. By conversion, and with due allowance for ion re-combination within the chamber, the current reading can be used to indicate the absorbed dose in gray. Most instruments in everyday use are, for technical reasons, 'sub-standard' chambers and are checked (or calibrated) against standard ionization chambers kept at the UK National Physical Laboratory, Teddington.

Since neutrons are uncharged they do not directly ionize matter and indirect means have to be used. Fast neutrons can be detected by causing them to interact with hydrogen atoms and it is the recoil protons (see p. 11) which cause the ionization. So, fast neutron monitors have a hydrogenous material, for example polythene, incorporated in their volume. To measure thermal neutrons, ionization chambers filled either with boron trifluoride (BF$_3$) gas or lined with a layer of boron are used. The capture of neutrons by the boron involves the emission of α particles and it is the latter that cause the detectable ionization currents.

The use of ionizations in air is the most important dosimetric method, but there are many other systems. Some of these are briefly listed below:

1. When lithium fluoride and many other crystals are heated after being exposed to radiation they emit light. The process is called 'thermo-

luminescence'. The absorption of radiation energy causes free electrons to be trapped in the lattice imperfections in the crystalline structure. These trapping levels lie between the valence and the conduction band; the electrons can remain trapped for a considerable period of time. If the crystal temperature is raised the electrons are excited from the trapping levels to the conduction band and then return to the valence level (the stable state) with the emission of light. The intensity of this thermoluminescence is a measure of the dose received by the crystals. Thermoluminescent dosimeters respond over a wide range of doses (10 μGy–1 kGy) of X-rays, γ rays, electrons and protons. Thermoluminescent devices may be used as powders in capsules or sachets or used as small rods or discs. The devices can be annealed after one exposure and used over and over again. Thermoluminescent dosimeters (TLDs) are widespread in laboratories for personnel monitoring.

2. Photographic emulsions consist of silver halide crystals or grains dispersed in gelatin. Radiation absorbed in an individual grain forms a latent image, and the chemical action of development reduces the grain to silver. The blacker the film is, as measured by densitometry, the more dose it has received. Film badges are the main form of personnel monitoring in radiation protection.

3. Scintillation detectors convert the radiation energy of ionizing particles into pulses of light. The size of a pulse is proportional to the energy deposited in the crystalline or liquid scintillant. Most scintillators have to be calibrated against other dosimetric devices because there is rarely a linear relationship between the dose and the pulse of light. The latter is detected by photomultipliers and recorded electronically.

4. Certain types of glasses and plastics become increasingly opaque with increasing radiation dose.

5. Solid state counters — some semi-conductors show an altered electrical conductivity during exposure to radiation. This method is analogous to the ionization chamber measurement, except that the current is flowing in the solid semi-conductor crystals and not in the gas of the ionization chamber.

6. Ferrous sulphate dosimetry is an accurate method of absolute dosimetry. Ferrous sulphate is converted to ferric sulphate when irradiated. Chemical titrations with potassium dichromate or ceric sulphate may be used to determine the amount of ferric ion present. Alternatively, the ferric ion yield may be assayed using an ultraviolet spectrophotometer, measurements being made at 304 nm. The yield is directly proportional to the dose received.

7. Finally, calorimetry, though not often used, is a direct method of measuring dose. It involves the measurement of the rise in temperature produced by radiation in an insulated mass of unknown thermal capacity.

The dose rate

Besides a knowledge of the absolute dose received by the biological material it is often necessary to know the rate at which the radiation is given (e.g. Gy h^{-1}, mGy s^{-1}).

The rate at which a given dose is delivered to a biological material can markedly alter the effect produced. This will be seen throughout this book (see Chapter 8). A dose of radiation, irrespective of the time it takes to be delivered, will produce an identical number of ionizations, and for a given quantity of ionizations we may expect a given amount of biological damage. However, it is often observed than at low dose rates a smaller biological response is observed that at higher dose rates, an observation that can be taken to mean that some repair of radiation damage is possible during irradiation at the very low dose rates (say 1 Gy day^{-1}) that is not possible at a high dose rate (1 Gy s^{-1}). On the other hand, some biological effects are not affected by the dose rate, which suggests that no modification of the amount of radiation damage is possible.

The splitting (or 'fractionation') of a single dose into two or more fractions separated by a time interval often results in less biological damage than is produced by a single large dose. For instance, 10 Gy given as one dose may kill almost 100 per cent of a population of cells, whereas two doses of 5 Gy each, given with a 24 hour interval between the first and second dose, may kill only 40 per cent of the cells. This indicates the possibility of repair of the damage produced by the first dose. The effects of splitting the radiation dose will be discussed more fully in Chapter 4.

The distribution of the dose in biological material

Although all ionizing radiation interacts with living matter in a similar way, different types of radiation differ in their effectiveness or efficiency in damaging a biological system. The 'relative biological effectiveness' (RBE) of a type of radiation is always expressed in relation to a dose of a standard type of radiation.

The most important factor that influences the RBE of a type of radiation is the distribution of the ionizations and excitations in its tracks. The idea of specific ionization was mentioned in the section on α particles (p. 7). In order to accommodate both the excitation and ionization events, the term linear energy transfer (LET) has been coined.

LET is expressed in terms of the mean energy released in keV per micrometre (μm) of the tissue traversed (keV μm^{-1}). As was seen in the case of specific ionization, the LET will be affected by the velocity and the charge of the ionizing particle. Alpha particles, neutrons and protons are high LET radiations and X- and γ rays and fast electrons are low LET radiations. Figure 2.4 shows a diagram of DNA with the relative distribution of ionization excitation events of three radiations superimposed upon it. Since the biological effectiveness of a particle is related to the amount of ionization and the distribution of that ionization in its tracks, particles with high LETs will be more damaging per unit of dose than low LET radiations. Alpha particles, protons and neutrons therefore have a higher RBE than X-rays, γ rays and electrons. The relative biological effectiveness of a radiation increases with increasing linear energy transfer. This increasing RBE with increasing LET does not hold at very high values of LET. At these high ionization densities, much more energy is deposited in the biological system than is necessary to produce an effect. Since much of the energy is 'wasted', the *relative* biological effectiveness falls. A general RBE/LET curve is shown in figure 1.12. A more detailed discussion of RBE and LET is given in Chapter 8.

To obtain the biologically effective dose for different types of radiation we must use the concept of 'dose equivalence'. This concept is most useful in radiation protection (see Chapter 11) where mixtures of radiations have to be considered because it takes into account the differences in the relative effectiveness of different radiations. The sievert (Sv) is the unit of dose equivalence and is numerically equal to a dose in grays multiplied by the quality factor (Q) of the radiation. Table 1.6 lists the quality factors that have been specified by the International Commission on Radiological Protection (ICRP) and the LET values that ICRP feel should carry these quality factors.

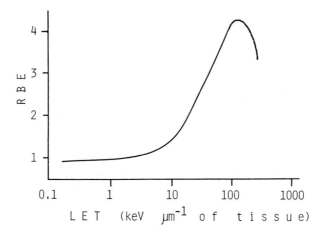

Figure 1.12. The general relationship between the relative biological effectiveness (RBE) of radiation and its linear energy transfer (LET)

Table 1.6.   The dependence of quality factor *(Q)* on LET for some radiations

LET in water (keV μm^{-1})	Quality factor	Radiation
$\leqslant 3.5$	1	⎫ X-rays, γ rays or
7	2	⎭ electrons
23	5	⎫ protons, neutrons
53	10	⎭
$\geqslant 175$	20	α particles, heavy recoil nuclei

Quality factors are based to some extent upon experimentally determined values of RBE of various radiations. The table also denotes the types of radiation most likely to be ascribed those quality factors. So for X- or γ rays a $Q = 1$ means that an exposure to 1 gray or 1 sievert are the same, while 10 Gy of fast neutrons ($Q = 10$) will cause the same damage as 100 Gy or 100 Sv of X-rays. The old unit of dose equivalence was the rem, and 1 sievert = 100 rem.

Chapter 2
The effect of radiation at the molecular and subcellular levels

2.1. Introduction

This chapter examines the effects of radiation at the biophysical, biochemical and subcellular levels, looking out for useful clues that might be a basis for interpreting radiobiological damage in tissues and whole organisms.

The effects of the physical deposition of radiation energy are quite well understood, and were dealt with briefly in Chapter 1. The biological effects of radiation will be described in the rest of this book. What is lacking is knowledge of the details of the links between the physical, chemical and biological effects of radiation.

Ionizations cause the majority of the immediate chemical changes in living material. This damage may be the *direct* result of an ionizing track or it may be due to the *indirect* action of free radicals (see p. 21). The chain of chemical reactions that results in damage to critical biological molecules may take as little as 10^{-6} second, while the final expression of biological damage may take hours, days or even tens of years.

Our knowledge of the molecular effects of radiation has tended to mirror our understanding of molecular biology in general. So, early radiation biochemists concentrated on the effects on proteins and particularly enzymes, but increasing attention in the 1960s and 1970s has been paid to the nucleic acids, especially deoxyribonucleic acid (DNA). Radiation biology and photobiology (UV effects) have opened up one of the most exciting fields of molecular biology—the phenomenon of molecular repair processes.

2.2. The effects of radiation on proteins

Proteins form the major organic basis of cytoplasm. They are complex molecules made up of chains of amino acids and may have any molecular weight from 10^3 to 10^6. The specific characteristics of a protein are determined by the sequence and nature of the amino acids in its chain (the primary structure) and by the complex foldings of its chain (the secondary and tertiary structures). Some proteins act as structural components in the cell while others act as the organic catalysts (enzymes) of the cell's biochemical reactions, and the latter have received most attention in radiation biology.

Studies have been made into both the physico-chemical and the biochemical effects of radiation on enzymes. The physico-chemical criteria of damage include a decrease in molecular weight due to fragmentation of polypeptide chains, changes in solubility, disorders of

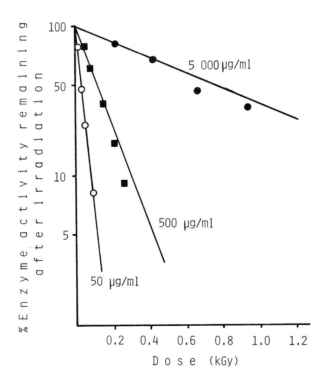

Figure 2.1.   The inactivation of the enzyme DNAase by various doses of radiation. The DNAase was irradiated at three different concentrations in solution
Source: S. Okada, 1957, *Archives of Biochemistry and Biophysics* **67**, 102; courtesy the author and Academic Press.

the secondary and tertiary structure, cross linkage and the formation of aggregates, as well as the destruction of amino acids in the chain. The biochemical criterion of damage is the loss of the ability of the enzyme to carry out its reaction.

As an illustration of the effect of radiation on an enzyme let us look at deoxyribonuclease (DNAase), the enzyme that acts by splitting deoxyribonucleic acid. Figure 2.1 shows the effect of different doses of radiation on the activity of the DNAase molecule *in vitro* at three different concentrations in solution. The first thing to notice is that as the dose of radiation increases so does the percentage of inactive molecules. In addition, the radiosensitivity, i.e. the response per unit of radiation dose, changes with its concentration in solution. What is in fact happening is that many more molecules are inactivated at low concentrations than at high concentrations. It seems that as the concentration of the enzyme falls, and the number of water molecules increases relative to the number of enzyme molecules, it becomes easier for the radiation to inactivate the enzyme molecules. This is a good illustration of the indirect action of radiation (see Chapter 1). At very high concentrations of the enzyme, the majority of the radiation effect will be due to the direct interaction of the radiation with the enzyme. In contrast, at low concentrations the damage done to the enzyme is caused predominantly by the diffusion of reactive free radicals of water.

Tens of gray are needed to cause an appreciable inactivation of the catalytic activity of an enzyme *in vitro*. Besides irradiating enzymes *in vitro* it is possible to irradiate cells and then to extract specific enzymes and test them for their catalytic activity. Once again, tens of gray are needed to give an appreciable *in vivo* effect. Such doses are an order of magnitude greater than those needed to cause significant damage to cells. For instance, an average of 1·5 Gy will kill two-thirds of a population of mammalian cells either *in vitro* or *in vivo* (see Chapters 3 and 4). It may be added that future technical development may allow the detection of enzyme changes at doses as low as 1–2 Gy.

However, it should be emphasized that enzymes are only sensitive to radiation when irradiated in dilute solutions and when no other proteins are present. Almost any inert protein, for example serum albumin, will protect enzymes against radiation damage, and in cells enzymes are always in close proximity to other proteins.

Besides the discrepancy between the doses necessary to affect enzymes and those needed to injure cells there are often problems associated with the interpretation of radiation changes in enzymes. For example, *in vivo* one can never be sure if the enzyme itself has been directly altered or whether the change in its activity is a result of radiation damage in other cell sites, such as the membranes. This suggestion that changes in enzymes *in vivo* may be secondary to other radiation damage is reinforced by the fact that it is generally true that larger doses are needed *in vitro* to produce a detectable change in an enzyme's function than is the case following irradiation *in vivo*.

The problem with enzymes is that there are so many of them, and most of them are being continuously produced, so that the loss of a sizeable fraction of them following a dose of radiation may be of no consequence to the cell. On the other hand, some enzymes might be present in very small quantities and at the same time be uniquely important to the cell so that the damage to them would effectively damage cells. No such enzymes have yet been found.

From the studies to date, it may be concluded that damage to recognized enzymes *in vitro* or *in vivo* usually requires doses much higher than those known to produce striking cellular changes and even cell death. This may be due to the insensitivity of present biochemical detection methods, the study of irrelevant enzymes, or to the fact that in the organized cell there is a target overridingly more sensitive to irradiation than the enzymes.

2.3. The effect of radiation on nucleic acids

The macromolecule of DNA consists of two spiral strands twisted around one another to form the now famous 'double helix' (figure 2.2). The strands are composed of a 'back-bone' of alternate phosphate and sugar groups and attached to each sugar group is a nitrogenous base. The two strands are held together by hydrogen bonds that run from one base to another. If the double helix were unwound, it would resemble a step-ladder (figure 2.2*b*), the upright of the ladder being the sugar–phosphate 'back-bone' (figures 2.2*c* and 2.2*d*) and the rungs being the combination of base pairs held together by hydrogen bonds (figure 2.2*e*). There are four bases, two pyrimidines (thymine and cytosine) and two purines (guanine and adenine). The number of adenine molecules in any specimen of DNA is always equal to the number of thymine molecules; and similarly, the number of cytosine molecules equals that of guanine. In fact, these pairs of bases always occur opposite one another on the helix (figures 2.2*b* and 2.2*c*). This makes the strands complementary to one another. There have always been problems associated with supercoiling and with the separation of the Watson–Crick double helix, and recently another model that does not involve intertwining of the strands has been suggested by Sasisekharan and his colleagues (see figure 7.1, p.111).

Watson and Crick suggested that the genetic information in the nucleus is held in the DNA molecules in the form of a linear sequence of bases—genes. The information is in the form of a four letter alphabet (the four bases), the 'words' of the code are of three letters each. Each triplet of adjacent bases, in messenger RNA (mRNA), codes for a particular amino acid. A specific sequence of bases corresponds therefore to a specific linear sequence of amino acids, i.e. to a specific protein molecule. So, genes at a molecular level are a linear array of nucleotide sequences (see also Chapter 7). Recent studies in molecular biology have revealed that in eukaryotic cells about 50 per cent of the base sequences are repetitive and only 50 per cent contain unique sequences of DNA. It also seems that these short repetitive DNA sequences are dispersed throughout the genome and alternate with the longer sections of unique sequence DNA. The function of repetitive DNA in gene organization and transcription is unknown. A further complication in the eukaryote genome is the finding that genes are not always made of a continuous sequence of codons that code for a complete protein. It seems that some genes are split and have non-coding base sequences (introns) interspersed between the sequences that do code for the protein (exons).

The amino-acid sequence that specifies a given protein is not built up directly on the DNA. Protein synthesis occurs in the cytoplasm and in order to transfer the DNA code to the site of synthesis, intermediary molecules of mRNA are used. A specific mRNA is an exact complementary copy of that part of the DNA that is to be translated into a protein. In the case of split genes the introns of the gene are excised so that only exons are transcribed by the mRNA. The mRNA detaches itself from the DNA and moves from the nucleus to the cytoplasmic site of protein synthesis, the ribosomes. Ribosomes are composed of protein and another species of RNA, ribosomal RNA (rRNA). On the ribosomes the mRNA acts as a mould (or template) for the synthesis of a specific protein. The individual amino acids required to build up the protein are brought to the ribosomal site by a third RNA, transfer RNA (tRNA). Transfer RNAs are coded for specific amino acids.

The three types of RNA, messenger, ribosomal and transfer, have a structure strictly comparable with a single strand of DNA, except that in RNA uracil replaces thymine. Cells are continuously manufacturing new proteins, with the result that the synthesis of the various kinds of RNA is a continuous process. RNA synthesis in fact only ceases for a very short period during the actual process of nuclear division (mitosis). In contrast, the process of DNA replication is not continuous and occurs at a definite part or phase of the cell division cycle (S phase). The other phases of the cell cycle are shown in figure 2.3. During replication (copying) the two strands of the RNA separate and two new strands are made using the complementary base pairings on the original two strands, thus forming two new DNA molecules. The two new molecules will eventually separate at mitosis (M), thus allowing the transmission of the identical genetic information to two daughter cells (see Chapter 7).

The phase prior to S phase is called G_1 and that just after it is known as

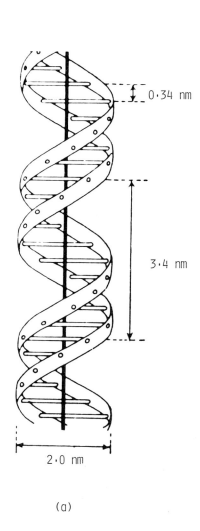

0·34 nm

3·4 nm

2·0 nm

(a)

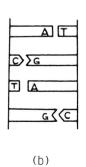

(b)

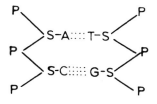

P   PHOSPHATE
S   SUGAR
A   ADENINE
G   GUANINE   ) PURINE BASES
C   CYTOSINE
T   THYMINE   ) PYRIMIDINE BASES
    H-BONDS

(c)

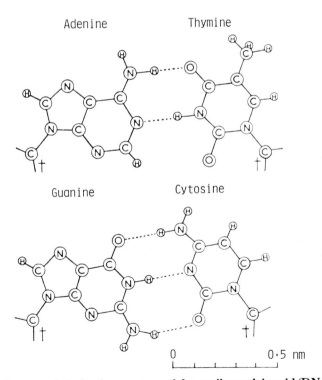

(d)

(e)

0   0·5 nm

Figure 2.2.   Diagram of the basic structure of deoxyribonucleic acid (DNA)

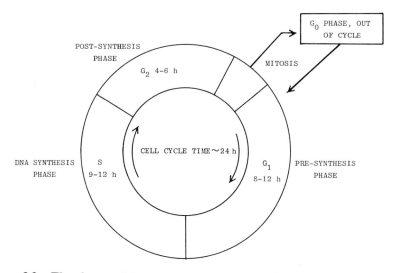

Figure 2.3. The phases of the mammalian nuclear cycle

G_2. The G stands for 'gap' but it must not be inferred from this that nothing happens in these phases as far as cell division cycle is concerned. During G_1 the cell is marshalling the complex biochemical apparatus of small precursor molecules necessary for the task of copying the DNA. These include the four nucleoside 5^1 triphosphate precursors and enzymes such as endonucleases, DNA polymerases and ligases. It has been shown that the replication of DNA occurs at between 10^3 and 10^4 sites (replicons) along the molecule at any one time during the S phase. The replication at these growth points proceeds simultaneously in two directions along the molecule until it links with the adjacent replicating points. S phase lasts about 6–12 hours in mammalian cells and is followed by G_2 before mitosis. In G_2 the DNA undergoes the complex coiling that leads to the formation of chromosomes which can be stained and observed at mitosis (see also figure 7.3). Occasionally subpopulations of cells can go into a phase called G_0 where they are essentially resting or 'out of cycle'. Mitosis and cell division are among the most 'radiation-sensitive' processes the cell performs, and much of this book will be concerned with the loss by cells of their ability to successfully complete mitosis and cell division following a dose of radiation. The killing effect of radiation in plants and animals can often be traced to the fact that radiation inhibits cell division (see Chapter 6). In contrast, it is generally true that cells that do not divide remain largely unaffected by doses of radiation that produce reproductive failure in dividing cells. Since DNA replication is essential for cell division, a vast amount of research has been concentrated on the effects of radiation on DNA and its synthesis. This work has been carried out at the following levels:

1. Major emphasis is placed upon radiation-induced molecular strand breaks and base damage and its repair.

2. Interest also centres upon radiation-induced deficiencies of DNA synthesis and division delay.

3. DNA is studied as the most likely 'target molecule' for cell death (see Chapter 3).

4. At the cytogenetic level damage to DNA is probably a crucial determinant of visible chromosome abnormalities (see Chapter 7).

5. At the genetic level the study of radiation-induced mutations obviously involves DNA changes (see Chapter 7).

6. Finally, there is the circumstantial evidence linking radiation damage to DNA with radiation carcinogenesis (Chapter 9), teratogenesis (Chapter 5) and even with radiation lifeshortening (Chapter 10).

2.4. Radiation-induced DNA damage and its repair

The study of the effect of radiation on DNA, and in particular the phenomenon of DNA repair, has produced great insights into molecular

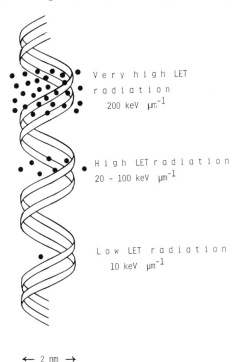

← 2 nm →

Figure 2.4. The relative distribution of ionization events for high and low LET radiations

biology, genetics and carcinogenesis. The most significant advances are the discovery of three major types of DNA damage and a number of important biochemical repair mechanisms. Figure 2.4 shows the relative distribution of ionization events for high and low LET radiation in relation to the size of the DNA molecule and figure 2.5 shows three of the major types of damage that such ionization events can produce: single and double strand breaks and base damage. Figures 2.7, 2.8 and 2.9 show three important repair pathways: excision repair; post replication repair; and proposed mechanisms for the repair of double strand breaks.

It is becoming apparent that DNA is subject to constant damage both from chemicals and from UV and ionizing radiation. A significant percentage of the energy of cells is spent synthesizing enzymes to repair and maintain the integrity of the genetic code, DNA. In the following sections we shall examine some of the recent findings in this burgeoning field. One note of caution must be introduced at the outset and that is the truism that those types of damage most measurable may not necessarily be the most significant.

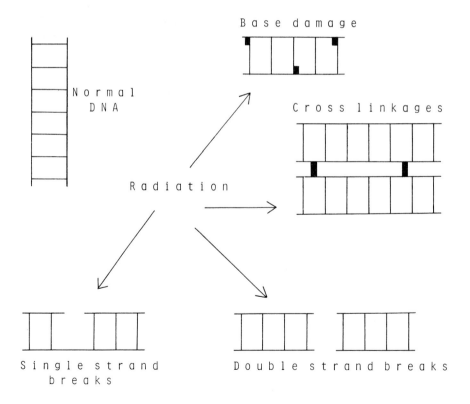

Figure 2.5. Three of the major types of radiation damage in DNA

Single strand breaks

Radiation-induced single strand breaks (SSBs) in DNA were first demonstrated using the 'alkaline sucrose gradient velocity sedimentation' technique. Extreme alkaline conditions (pH 12) are used to split the DNA lengthwise into its constituent strands. The cells are layered onto a specially prepared sucrose solution and centrifuged at high speeds. The DNA strands sediment at specific velocities which can be related to their molecular weight. If the DNA has been irradiated and a number of breaks induced, then the lengths of the strands (and so their molecular weight) will be less than the unirradiated control DNA. Single strand breaks can also be accurately determined at low doses (<1 Gy) using the recent alkaline elution method. Figure 2.6 shows the relationship between the dose and the amount of DNA retained on a polycarbonate filter. As the dose increases the average molecular weight of the DNA pieces falls and so less and less is retained by the filter.

It has been shown that the number of SSBs is linearly related to the dose of radiation over a very wide dose range, from less than 0·2 Gy to 60 000 Gy. This implies that however small the dose there will be some SSBs. The efficiency of induction of SSBs varies with a number of

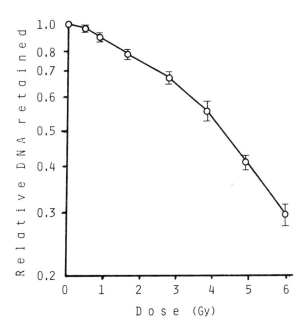

Figure 2.6.   The fraction of DNA retained by a filter as a function of dose
Source: M. D. Mills and R. E. Meyn, 1981, *Radiation Research* 87, 319; courtesy the authors and Academic Press.

biochemical factors, but recent measurements give values for the average energy required of between 10 and 20 eV per break for low LET radiation. Under normal conditions a significant proportion of SSBs are induced via a mechanism involving the OH• radicals of water. This has been proved using chemicals known as 'OH• radical scavengers'. The number of SSBs induced by radiation in oxygenated mammalian cells is 3–4 times that found in cells irradiated under hypoxic conditions (see Chapter 8).

The repair of SSBs in mammalian cells is very rapid and efficient. It probably occurs by the mechanism called 'excision repair' (see figure 2.7). This involves the excision of the length of nucleotide strand containing the defective piece of DNA and uses the complementary (undamaged)

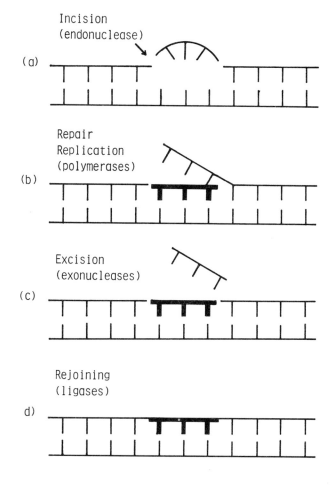

Incision
(endonuclease)

(a)

Repair
Replication
(polymerases)

(b)

Excision
(exonucleases)

(c)

Rejoining
(ligases)

d)

Figure 2.7. Diagram of the excision repair mechanism for the removal of such radiation induced lesions as DNA base damage and single strand breaks

single strand as the template for the resynthesis of a new length of DNA. The process is enzymatically controlled and temperature dependent, with no repair occurring at 0 °C. The first step in the process is the recognition of the site of the distortion or disruption when an incision is made by endonuclease enzymes (figure 2.7a). The incision is followed by the complete excision of the lesion and sometimes a wide area around it and this involves endonucleases. The gap produced is then filled with new nucleotide bases by the action of DNA polymerases using the opposite and undamaged strands as the guide (figure 2.7c). When this 'repair replication' is complete the new section of DNA is linked to the intact DNA by enzymes called ligases (figure 2.7d). In mammalian cells the rate

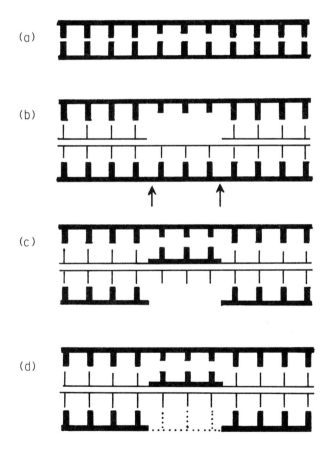

Figure 2.8. A general model for post replication UV radiation induced DNA damage
⬆ = endonuclease incision. The parental DNA strands are the heavy lines, the newly synthesized strands are the feint lines, the dotted lines represent repair replication. The damage itself is not repaired but bypassed.

of excision repair is exponential, with half of any radiation-induced SSBs being repaired within 15 minutes. Since most breaks are repaired even in lethally irradiated cells it is felt that they are much less important determinants of cell death than either double strand breaks or base damage. Unrepaired single strand breaks can, of course, take part in the formation of double strand breaks.

Figure 2.8 shows the general model for post-replication repair of radiation induced DNA damage. This model, like much of the DNA repair story, was first elucidated by photobiologists working with UV-damaged micro-organisms. In the model, DNA synthesis proceeds past the point of damage and so leaves a gap in the newly formed daughter strand (figure 2.8*b*). The gap is then filled with a piece of parental strand by a recombinational process (figure 2.8*c*). The gap so formed in the parent strand is then filled by repair synthesis analogous to that in figures 2.7*b*, *c* and *d*. It is important to notice that the damage itself is not actually repaired but is bypassed and the lost genetic information is supplied by redundant information within the cell. There is little evidence that this type of post-replication repair plays a role in the repair of ionizing radiation damage in mammalian cells.

Double strand breaks

As figure 2.5 shows, double strand breaks (DSBs) can be formed either by a single ionizing event or by the coincidence of random single strand breaks on the complementary strands. The measurement of DSBs involves neutral gradient sedimentation methods. The neutral biochemical extraction conditions do not split the DNA molecule into single strands and the number of DSBs is measured by a decrease in the average molecular weight of the pieces of DNA. However, the technique is only capable of distinguishing between relatively small pieces of DNA (<1–3×10^9 dalton molecular weight) and so relatively large radiation doses have to be given (50–100 Gy) to chop the DNA into small enough pieces to measure. The sedimentation method has a number of disadvantages in mammalian cells, most associated with anomalous sedimentation patterns due to DNA complexing with membranes and other 'contaminants'. A further method has very recently been developed that can measure both DNA breakage and base damage. This 'unwinding method' as it is called has great potential since it can be used at much lower doses (~10 Gy) than the older methods.

The relationship between dose and the number of DSBs induced remains in doubt at least in mammalian cells. Some workers find a simple linear response while others have shown that only the initial part of the curve at the lowest doses is linear and at higher doses the number of

DSBs increases with some power of the dose, i.e. the dose effect curve is 'linear quadratic'. Such a response indicates that DSBs may be produced by the passage of one ionizing event or as a result of two independent single strand breaks. The significance of 'linear v. linear quadratic dose response curves' will become clear when we discuss cell killing (Chapter 3) and radiation risks (Chapters 8, 9 and 11).

The repair of DSBs is conceptually difficult to imagine. The nucleotide base sequence may not be available on either strand and there may be no continuity between the two 'free ends'. These difficulties have led to the conclusion that perhaps DSBs were irreparable and therefore were invariably lethal lesions. However, there is now irrefutable evidence in X-rayed bacteria, yeasts and mammalian cells that DSBs can be repaired, but the methods used can only detect whether the free ends of a broken DNA molecule have joined together and cannot indicate whether the original base pairing of the genetic code has been re-established. The methods cannot detect the presence of what is called 'misrepair' or 'error prone' repair which might be a cause of significant genetic damage (see Chapter 7).

Figure 2.9 shows the details of a theoretical model that has been proposed by Resnick to explain how radiation-induced DNA double-strand breaks might be repaired. Figure 2.9a shows the double strand break and figure 2.9b shows the enzymatic excision of parts of a single strand for each of the broken ends, leaving single-stranded pieces. In the model these single strands are then imagined as associating with homologous and undamaged strands in such a way as to allow recombinations to occur in much the same way as we saw in figure 2.7 for post-replication repair. The undamaged strands have to be cut or 'nicked' by an endonuclease in two places as shown by the arrows in figure 2.9b. Figures 2.9c and d show how a reciprocal exchange of the DNA helices could occur. The recombination will leave two small and two larger single-strand gaps. The small one can be sealed by ligase enzymes and the larger gaps repaired by the repair replication shown in figures 2.7b, c and d. These final repairs are shown as the dotted lines in figure 2.9e. It is important to bear in mind that, unlike excision and post-replication repair, there is as yet no experimental evidence to prove or disprove the recombinational process just outlined for the repair of double breaks in DNA.

DNA base damage.

Radiation damage to the purine and pyrimidine bases of DNA was first recognized as important in bacteria. Recently a number of highly sensitive tests have been developed to actually measure such lesions, especially thymine damage. Such damage is linearly related to dose and is thought to arise via interaction with aqueous free radicals such as the OH˙. Radiation-induced thymine base damage is more frequent in mammalian cells than single strand breaks. The excision repair mechanism (see figure 2.7) seems to be responsible for the rapid and efficient removal of damaged bases in both bacteria and mammalian cells. In prokaryotes unrepaired base damage is an important determinant of survival and there is increasing evidence that similar damage may be important in higher cells. For example, Ataxia telangiectasia (AT) is a rare human genetic disease that affects 25 in a million births. The disease involves many systems including the skin, the nervous and the immune systems. About 10 per cent of patients develop cancer before the age of 20 years, mostly leukaemias and Hodgkin's disease. An interesting aspect of AT patients is their extreme radiosensitivity and the reduced ability of their cells to repair potentially lethal damage (see also p. 75). The most likely cause of the radiosensitivity is their failure to repair some types of base damage. AT cells are normal for excision repair and for the rejoining of both single and double strand breaks in DNA.

It is now apparent that the repair of chemical or radiation damage to DNA is a central cellular function. There are broadly three types of repair. First, there is 'error free' repair, mainly excision repair, that causes no lethality and no mutations. This involves the removal of damaged DNA and its replacement with undamaged nucleotides thus restoring normal DNA function (see figure 2.7). Second, there is 'error prone' repair which may produce non-lethal or lethal mutations. Here the damage is not itself immediately repaired but is bypassed during DNA replication leaving gaps in the daughter strands. The missing genetic material is supplied in the 'post-replication repair' step via a recombination process from the parental DNA strand (see figure 2.8). And third, there is incomplete repair where the continuity of the DNA strands is not re-established. Such non-repair is lethal but not necessarily mutagenic, at least in bacteria. The discovery of these multiple DNA repair pathways, and there are more than those noted above, has profoundly influenced radiation biology. It has also had reverberations in biology in general and in the study of genetics and cancer in particular, which we will discuss later in this book (Chapters 7 and 9).

2.5. Radiation effects on DNA synthesis and division delay

As we shall see in Chapter 3 a dose of 1–1·5 Gy of X-rays will kill approximately two-thirds of a population of mammalian cells. After such a dose all the cells will have sustained much base damage and have about 1000 strand breaks in their DNA of which about 50 will be double breaks. It follows that the one-third survivors have repaired large numbers of such lesions or are capable of survival with a genome peppered with random double strand breaks. Surviving cells are certainly not undamaged cells. This leads to the crucial question of whether DNA damage is the main determinant of cell death, and this will be discussed more fully in Chapter 3.

It seems reasonable to suggest that such radiation effects as mitotic

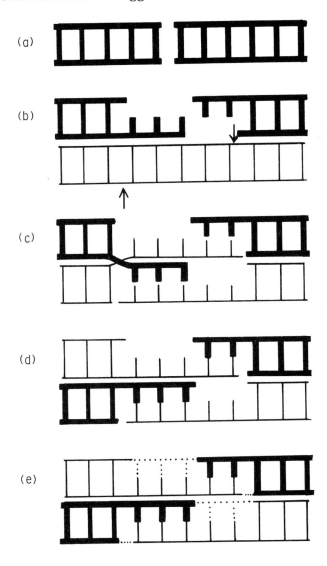

Figure 2.9.   A proposed model for the repair of DNA double strand breaks
↑ =endonuclease incisions.

delay, the delay in DNA synthesis and the overall reduction in the amount of DNA synthesis are a result of DNA base damage and strand breakage. However, there is little evidence to support this assumption. The term mitotic or division delay refers to a transitory hold up of cells at a point in G_2. Irradiated cells take up to 1 hour per gray longer (see figure 2.10) to reach their first post-irradiation mitosis than unirradiated cells.

Division delay cannot be induced unless the nucleus or perinuclear region is irradiated. A recent series of experiments used iodine-125 (^{125}I) attached to pyrimidine bases, which enabled the ^{125}I to be incorporated into the DNA molecule. The ^{125}I decays by emitting an extremely localized shower of electrons that travel only about 25nm. The conclusion of the experiments was that the subcellular target for division delay is not the DNA but is most likely the nuclear membrane. The G_2 blockage seems to be due to interference with the chromatin condensation process that is essential for the formation of mitotic chromosomes and which occurs while the DNA is in close association with the nuclear envelope.

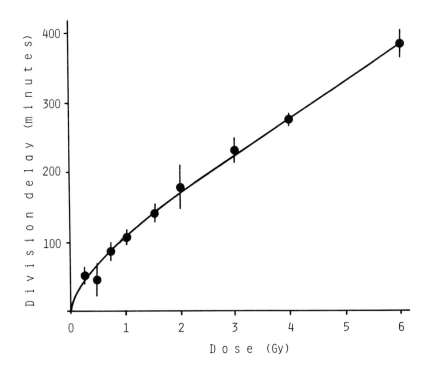

Figure 2.10. **The delay of cells entering mitosis (due to a G_2 block) as a function of dose**
Source: B. F. Kimmler, D. B. Leepes and M. H. Scheiderman, 1981, *Radiation Research* **85**, 276; courtesy the authors and Academic Press.

It is difficult to summarize the experiments on the post-irradiation depression in DNA synthesis as measured by the radioactively labelled precursor of DNA. The degree of depression is not only dose-dependent but markedly depends on the phase of the cycle at which the radiation is given. For example, in one cell line 5 Gy given to cells in mitosis caused a 70 per cent reduction in synthesis, whilst 5 Gy given to G_2 cells produced only a 20 per cent deficiency in the next S phase. Other cell lines have differing sensitivities. However, it seems true to say that the most sensitive phases for radiation-induced defective synthesis (mitosis and the G_1/S border) are also the most sensitive for cell killing. Once in synthesis cells are often least radiosensitive both for cell killing and the depression of DNA synthesis. The molecular mechanisms of defective DNA synthesis remain obscure.

Deficient DNA synthesis should lead to deficient RNA synthesis and so to deficits of essential proteins. However, there are few studies of the effect of radiation on either RNA or protein synthesis. It appears that RNA synthesis is less sensitive to delay and suppression than DNA synthesis. One recent study indicates a slight increase in RNA synthesis after about 7 Gy, while another study reports a post-irradiation depression of RNA synthesis in cells given 5 Gy! The quantitation of both RNA and protein synthesis are more difficult than DNA and there has been virtually no research into the relative radiosensitivities of mRNA, rRNA or tRNA synthesis. The effect of radiation on RNA and protein synthesis are generally unimportant in radiation biology.

2.6. Nuclear damage versus cytoplasmic damage following radiation

Some early radiobiological experiments were designed to distinguish whether the nucleus or the cytoplasm is more affected by radiation and whether it is possible to kill cells if the cytoplasm only or nucleus only is irradiated. The answer is two-fold.

Experiments show firstly that the nucleus is the more important site for the lethal effect of radiation and secondly that only if the dose to the cytoplasm is large will it cause cell death. In other words, the nucleus is more radiosensitive than the cytoplasm for radiation cell killing. But what type of experiments led to this conclusion?

1. Many experiments involved techniques that included the nucleus in, or excluded the nucleus from irradiation. A tiny speck of a radio-isotope, such as polonium-210, can give a beam of α particles that travels only 40 μm in tissue. With such microbeams it is possible to irradiate the nucleus and leave the cytoplasm almost completely unirradiated, or vice versa.

2. Another technique involves the removal of nuclei from cells by

micro-dissection. The empty cytoplasm can be irradiated and the unirradiated nuclei re-implanted, or an irradiated nucleus can be implanted into unirradiated cytoplasm.

Both these methods showed the nucleus to be up to 100 times more sensitive to a given dose of radiation than the cytoplasm when post-irradiation cell division and survival were the endpoints tested.

3. An ingenious experiment on cells using internally deposited radioactive bases of RNA and DNA also points to the greater radio-sensitivity of the nucleus. Thymine, one of the DNA bases, is found almost exclusively in the chromosomes in the nucleus, while uridine as one of the RNA bases is found throughout the nucleus and cytoplasm. Both of these bases can be labelled with tritium (^{3}H), the radio-isotope of hydrogen that emits β particles. Once inside the cell the β rays from the ^{3}H-uridine will irradiate all parts of the cell, whereas the β rays from the ^{3}H in the ^{3}H-thymidine† will deposit all their energy in the nucleus. In fact, since the β particles from ^{3}H are only 18 keV particles, they travel no more than 1–2 μm in tissue and most of the energy of the ^{3}H in the ^{3}H-thymidine will be dissipated within the chromosomes themselves. The experiment involved the use of combinations of radioactive and non-radioactive thymidine and uridine and showed that the exposure of the cytoplasm to ^{3}H β rays did not contribute to the observed cell killing. All the cells which were killed died as a result of β rays depositing their energy inside the nucleus. Other radiotracer studies comparing the lethal effects on cells of tritiated water with tritiated thymidine showed that one thousand times more radiation was needed from the water compared with the thymidine to give the same killing effect. Similarly, the radiation from ^{125}I-labelled compounds incorporated into DNA is 200–300 times more effective than ^{125}I compounds that attach themselves to the outer cell membrane and merely irradiate the cytoplasm. This type of work not only implicates the nucleus as a critical site for damage but pinpoints the chromosomes as critical entities for radiation damage.

4. Such factors as the volume of the nucleus, the volume of the interphase chromosomes and the DNA content per chromosome or per nucleus are often of great value in assessing the likely radio-sensitivity of plant, insect, amphibian and even mammalian species and cells. The most useful parameter is the interphase chromosome volume (ICV), and since DNA content is directly related to ICV this strongly implicates DNA as a 'target'. A detailed discussion of the evidence for and against DNA and the chromosomes as the critical target for much radiation damage appears in the next chapter.

---

†Thymidine, or more correctly thymine deoxyribonucleoside (TdR), is a nucleoside formed by the condensation of thymine with a deoxypentose sugar. It is readily taken up by cells and acts as a specific precursor for DNA.

The studies outlined above and many others have led radiobiologists to conclude that although it is possible to kill cells and to produce other kinds of damage by irradiation of the cytoplasm, it is much easier to produce the same effect by irradiation of the nucleus. This conclusion is perhaps to be expected because many of the criteria of radiobiological damage are intimately related to the function of the nucleus. They include chromosome damage, delays in the onset of mitosis, the cessation of mitosis and cell division (reproductive death), and the inhibition of DNA synthesis. Nor is the pre-eminence of the nuclear effects surprising when it is remembered that it is currently believed in biology that the majority of information in a cell resides in the nucleus—as the genetic code in the DNA. If the nucleus does contain nearly all the necessary information for the cell processes, it is to be expected that any damage to it will be of primary importance, whereas damage to the cytoplasm may be of secondary importance.

2.7. *Radiation damage to the cell membranes*

The cell may be described as a complex of interconnected membranes, since there are suggestions that the outer cell membrane, the endoplasmic reticulum, and the mitochrondrial, lysosomal and nuclear membranes and the Golgi body are all in intimate connection.

One role of the outer cell membrane is to regulate the exchange of material between the inside and the outside of the cell. However, very little is known about the dynamics of the structure and function of cell membranes or about the details of the regulatory roles they might play.

There are several cases of radiation affecting the outer cell membranes. For example, doses in the tens of gray range cause a reduced amplitude and a decreased conductivity of the nerve impulse in isolated peripheral nerves in adult animals. The transmission of the nerve impulse is known to be a result of the differential diffusion of sodium and potassium ions across the membrane of the axon. Such radiation-induced changes in the electrical activity of nerves indicate an increase in passive ion permeability of the nerve axon. Behavioural and functional effects in the central nervous system of adult animals have been shown after doses as low as 0·5 Gy. It is not known whether such effects are due to primary radiation damage in the nervous tissue or indirectly to the release of toxins from other damaged tissues, such as the endocrine, cardiovascular or gastrointestinal systems.

Radiation-induced changes in the properties of cell membranes have also been shown in the epithelial cells of the mammalian intestine. Following a 30 Gy dose, microvilli on the surface of these cells swell and eventually the whole cell becomes bloated, which is interpreted as being a result of the loss of the cell membrane's ability to regulate electolytes. Similar changes in permeability of the outer cell membrane have been reported in red blood cells, muscle cells and yeast after radiation.

Radiation damage to the endoplasmic reticulum

The endoplasmic reticulum (ER) is a system of membranes that runs throughout the cytoplasm and to which many enzyme systems are attached. The enzymes require an intact membrane for full activity and whenever the membrane is damaged, for example by radiation, enzyme activity will be impaired even though the enzymes may be unaffected. Ribosomes are also attached to the ER system and it is on ribosomes that the mRNA code is translated into the amino-acid sequence, giving a specific protein molecule. There are almost no radiobiological data that show that damage to the ER leads to a reduction of protein synthesis.

Radiation damage to the lysosomes

Several years ago an 'enzyme release hypothesis' was proposed suggesting that following radiation-induced membrane damage, the unplanned release of enzymes might occur inside a cell and cause its death. This hypothesis has never been satisfactorily proved, but it has become more plausible recently with the demonstration of the subcellular organelle known as the lysosome. Lysosomes contain a collection of enzymes, some of which play a role in breaking down ingested food and other foreign particles inside the cell. The 'unplanned release' of such catabolic enzymes following radiation membrane damage might be fatal to the cell.

Recent research has shown that lysosomal activation is an important early event after radiation. Such effects as the increased permeability of lysosomal membranes and a five to sixfold increase in lysosomal enzyme activity can be detected in spleen and thymus cells within 1–3 hours of 8·5 Gy of ^{60}Co γ rays. The 'free' lysosomal enzymes are capable of damaging proteins, mucopolysaccharides and nucleic acids. However, much more research is needed before a conclusion can be reached as to the definitive role of lysosomes in radiation biology.

Radiation damage to the mitochondria

Mitochondria are the primary sites for oxidative phosphorylation that results in the storage of energy for the cell in the form of high-energy chemical bonds. It is believed that the mitochondrial membranes must be intact for their correct functioning—the enzymes being arranged in a definite sequence on these membranes. One imagines that if radiation were to knock holes in the membranes, a step in the oxidative process might be deleted.

Experiments have shown that both the structure and the function of mitochondria are markedly affected by moderate doses of radiation. In liver cells a 10 Gy dose causes mitochondria to become globular and to fragment, and this is accompanied by a 50 per cent reduction in the amount of oxidative phosphorylation. The inhibition of oxidative phosphorylation has also been detected in the thymus, a particularly radiosensitive tissue, at doses as low as 0·25 Gy.

The significance of radiation damage to membranes, and to nuclear membranes in particular, is not fully understood. Nevertheless, there is an influential school of thought that emphasizes the close association between nuclear membranes and DNA and infers from it a major role for membrances as a prerequisite for many radiation effects in both prokaryote and eukaryote cells.

2.8. Summary

In the introductory remarks of this chapter it was said that studies at the biochemical and biophysical levels might provide a link between the physico-chemical and the biological effects of radiation.

At present damage to DNA appears to be the most likely link. This chapter outlined some of the current data on and approaches to DNA base damage and strand breakage and details of a number of known repair mechanisms. As we shall see throughout this book these molecular mechanisms are beginning to have an impact on such effects as radiation genetics, cell killing and carcinogenesis.

The importance of much of radiation biochemistry to protein (enzymes) or RNA is difficult to assess, but it seems to be less critical than to DNA damage. The role of membranes in radiation biology remains enigmatic but the close cohesion between DNA and the inner membranes of the nucleus means that membranes cannot be ignored as significant radio-biological 'targets'. Finally, it must always be remembered when we are discussing DNA that those lesions which are most readily measurable are not necessarily the most significant.

Chapter 3
Cellular effects of radiation (I)

3.1. Cell killing

Viruses, bacteria and plant and animal cells can be killed by ionizing radiation. As the dose of radiation increases there is an increase in the proportion of cells killed.

In the case of mammalian cells, at very high doses (tens of gray) radiation can cause the rapid cessation of cellular metabolism and cellular disintegration. This type of death, often termed 'non-mitotic death' or 'interphase death', is seen in such non-dividing or rarely dividing cells as those of the adult liver, muscle and kidney and in neurones. However, radiation at much lower doses can also 'kill' cells by inhibiting their ability to divide and this failure to proliferate is termed 'reproductive death', which is defined as "the loss of a cell's ability to undergo unlimited cell division". Consequently, cells capable of a limited number of post-irradiation divisions that produce sterile progeny are defined as killed even though morphologically, physiologically and biochemically they may appear normal. This term reproductive death will, by definition, also cover dividing cells that may degenerate after moderate doses (a few gray) even without attempting to divide, since they have truly lost their reproductive ability.

It is important to stress that in this chapter and throughout the book the terms radiation 'cell killing' and 'cell death' refer to the loss of unlimited reproductive capacity. It will be apparent that the term reproductive death is not applicable to cells that rarely if ever divide. The definition of cell death as the loss of proliferative ability might seem excessively narrow to most biologists. And indeed it is compared to a pathologist's definition of cell death which involves the visible disintegration of the cell. A radiation killed cell may not exhibit any signs of damage *until* it attempts to divide and such a 'dead cell' may not die

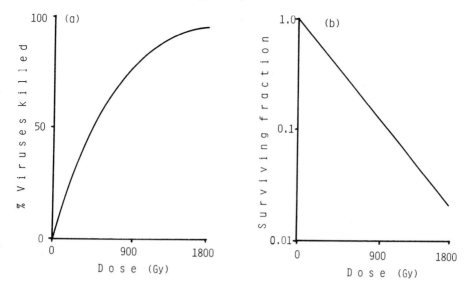

Figure 3.1. (*a*) **A linear plot of the percentage of viruses surviving against the radiation dose in gray;** (*b*) **A semi-logarithmic plot of the percentage of viruses surviving against the radiation dose in gray**

for weeks or months after exposure to radiation. Radiologists justify their use of a proliferative criterion of death on at least two grounds. Firstly, they have always had close links with radiotherapists and together they have attempted to optimise cancer treatment. The inhibition of the proliferative activity of the cells of a cancer is tantamount to its cure. Radiation is a most effective anti-mitotic agent. Secondly, as we shall see there is ample evidence that many of the acute and longterm effects of radiation are a result of the post-radiation mitotic death of cells.

The graphical presentation of the fraction of cells surviving against the dose of radiation received is known as a survival curve. This chapter is about survival curves and the theory that has been proposed to explain their shapes.

3.2. Simple survival curves

Figure 3.1*a* shows the percentage of viruses killed by different doses of radiation. As the dose increases, the proportion of viruses killed approaches, but never reaches, 100 per cent. The curve in figure 3.1*a* can be transformed into a straight line by plotting the surviving percentage on a logarithmic scale against the dose on a linear scale (figure 3.1*b*). This semi-logarithmic plot shows that the relationship between survival and dose for viruses is exponential, i.e. any unit increase of dose produces a corresponding fractional decrease in survival. This type of simple exponential relationship has been demonstrated for the radiation inactivation of certain biological molecules such as enzymes, and for the survival of viruses, some bacteria and in certain cases mammalian cells. However, mammalian cell survival curves are generally a little more complex, as we shall see later.

In 1924 it was proposed that direct hits on radiation-sensitive targets could explain radiation inactivation of molecules and cells. This 'target theory' was developed by Lea in the 1940s and still forms the basis of much radiobiological thought. Target theory attempts to relate two sets of facts, the physical facts about the absorption of radiation with the observed facts of the exponential shape of survival curves.

The fact that survival curves are exponential suggests the occurrence of a random process of killing. In Chapter 1 we saw that radiation energy is dissipated in living matter by the random processes of ionization and excitation and it is these events that are considered to be the cause of biological damage. Target theory proposes that inside cells there are certain critical molecules, critical sites or 'targets' that must be inactivated (or 'hit') by radiation if the cell is to be killed. If no hit is received in the target the cell will survive. Exponential survival curves of the type in figure 3.1*b* can be derived in theory if a single hit in a single target is sufficient to cause cell death. Such single-hit kinetics can apply to molecules, viruses, bateria and, under some conditions, mammalian cells.

The exponential survival curve in figure 3.1*b* may be described by the equation

$$N = N_0 e^{-D/D_0}$$

where N_0 is the initial number of viruses; N is the number of viruses that remain unaffected after a dose of radiation D; D_0 is a constant and gives an indication of the slope of the line. If one puts D_0 equal to D in the above equation this results in the term e^{-1} which equals 0·37 so that

$$N = 0·37 \, N_0$$

This means that D_0 is the dose required to reduce the population of viruses to 37 per cent of its initial value. In fact, from the binomial distribution law, D_0 is that dose required *on average* to put one inactivating event (or hit) in each of the viruses. This seeming contradiction in the definition of D_0 is a result of the random nature of the ionization events. Due to this randomness some viruses will receive more than one hit and some will receive no hits. *On average*, however, following a dose of D_0, approximately 63 per cent of the viruses will be inactivated and approximately 37 per cent will survive. The D_0 for viruses varies with the strain but the median value is 500 Gy.

Using the mathematics of survival curves, we can under some circumstances calculate the actual volume of the target. These conditions are:

(*a*) that the inactivation should be solely the result of radiation and not due to any biological influences;

(*b*) that the inactivation is the result of direct ionization in the biological target and not the indirect action of the diffusion of free radicals into the target volume (p. 21);

(*c*) that there should be no dose-rate effect which might indicate biological repair of some of the targets, etc. that would alter the slope or gradient of the survival curve.

It will be obvious that these conditions are very rarely operative, and in fact calculation of the molecular weight or size of the target has only been possible in the case of some large molecules and viruses, irradiated in the dry state (i.e. with no aqueous free radicals).

3.3. Bacterial cell survival curves

Different bacterial strains have different types of survival curves. Three types are seen in figure 3.2.

All three curves in figure 3.2 are plotted semi-logarithmically, with the surviving fraction plotted on a logarithmic scale against dose on a linear

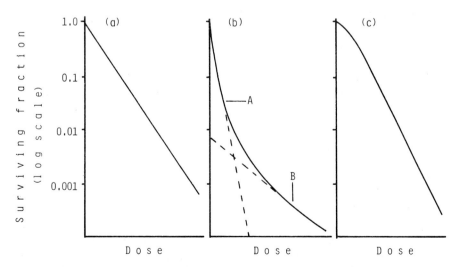

Figure 3.2. **Three types of bacterial cell survival curves.** (*a*) **Single-hit exponential survival curve;** (*b*) **Biphasic survival curve;** (*c*) **Multi-event survival curve**

scale. Figure 3.2*a* is a simple exponential curve, similar to that in figure 3.1*b* for viral inactivation. However, the D_0, which is the parameter of sensitivity measured from the slope of the exponential line, depends on the bacterial strain, and is generally between 10 and 250 Gy. Bacteria are therefore more sensitive to radiation than viruses.

Figure 3.2*b* is a biphasic survival curve. This curve is in fact the result of the irradiation of two populations of bacteria, each having a different radiosensitivity (i.e. different D_0). The dotted lines emphasize the dual nature of the curve. Such a biphasic curve might result from the irradiation of a mixture of rapidly dividing cells which are more sensitive (slope A) and stationary phase cells, which are less sensitive (slope B). Alternatively, the biphasic curve may be the result of two different strains of bacteria with different sensitivities; slope A corresponding to the radiosensitive strain and slope B to the resistant strain.

Finally, figure 3.2*c* represents a survival curve which has a 'shoulder' at low doses and becomes exponential only at higher doses. The D_0 of such a curve must be measured only on the straight-line, exponential portion of the curve. This type of inactivation curve is typical of mammalian cells and will be discussed in the next section.

3.4. Mammalian cell survival curves

In 1956 a breakthrough was achieved in radiobiology, when a radiation survival curve was obtained *in vitro* for a population of mammalian cells. The method, illustrated in figure 3.3, is to pipette a known number of single cells, suspended in a liquid nutritive medium, onto a petri dish. The cells attach themselves to the bottom of the dish, which is incubated at 37°C. At the end of 10 days or so, small isolated colonies or clones of cells are seen in the dish which are the result of individual cells having undergone a series of cell divisions. If the single cells are irradiated soon after they are plated onto the petri dishes some of them will be killed, resulting in fewer macroscopic colonies developing. The ability to undergo five or more cell divisions following irradiation is used as an indication of cell survival, since it is known that cells capable of such successful divisions are capable of almost indefinite cell multiplication. Conversely, cell death is indicated by a cell's inability to proliferate and give rise to a visible, macroscopic colony of some 32–64 cells, which represent some 5 or 6 successive doublings. But what is the fate of the radiation-killed cells? Cells of the irradiated population which do not survive may behave in one of several ways:

(*a*) Most cells undergo a few divisions, but less than the required 5 or 6, and then stop. This produces what are called abortive colonies. For example, after 4 Gy, 90 per cent of mouse 'L cells' are able to complete the

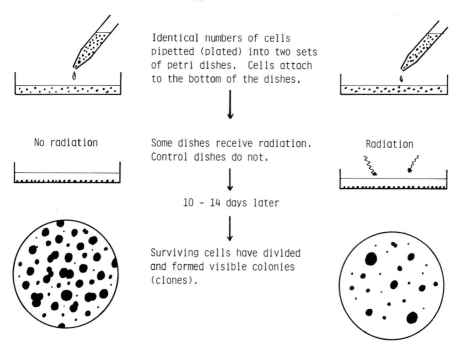

Figure 3.3. Diagram of the cloning technique of Puck and Marcus for the determination *in vitro* of mammalian cell survival following radiation
Source: T. T. Puck and P. T. Marcus, 1956, *Journal of Experimental Medicine* **103**, 653.
Not all the control (unirradiated) cells form clones so that

$$\% \text{ Plating efficiency (PE)} = \frac{\text{mean number colonies/dish}}{\text{number of cells plated/dish}} \times 100$$

Surviving fraction after dose D =

$$\frac{\text{mean number of clones after dose D/dish}}{\text{mean number of cells plated/dish}} \times \frac{100}{\text{PE}}$$

first division but 75 per cent fail to complete a second and die. Most such cell killing occurs within 2 or 3 divisions.

(*b*) Some of the cells that lose their ability to divide either immediately or after a limited number of divisions become enlarged and are known as 'giant cells'. They have retained their metabolic abilities and the synthesis of DNA, RNA and protein continues normally, the cells getting larger and larger. In some cell lines giant cells are formed by the fusion of daughter cells that are held together by chromatin strands at mitosis as bipolar or even multipolar configurations.

(*c*) Some cells undergo apparently normal cell divisions, but at a much slower rate than the controls. The result is that the 'slow-growing colonies' are too small to count as survivors at the end of the experimental period, although in time they might complete 5 or 6 divisions.

(*d*) Finally, some cells degenerate, lyse and become detached from the

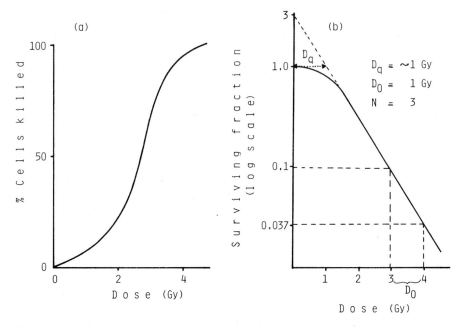

Figure 3.4.   The generalized survival curve for radiation-induced loss of reproductive capacity ('cell death') in mammalian cells. (*a*) **A linear plot of the percentage of cells killed against dose;** (*b*) **A semi-logarithmic plot of curve** (*a*) **with the fraction of surviving cells agains dose**

petri dish in attempting their first or subsequent post-irradiation divisions. The pattern of cell death varies from cell line to cell line. Some cells show a predominance of mitotic deaths while others may show interphase death or giant cell formation. The range of possible biological effects among irradiated cells produces a distribution of colony sizes which broadens with increasing dose. This is in contrast with the uniformity of colony size produced by unirradiated control cells.

Despite the somewhat arbitrary nature of the radiobiological definition of cell death as the 'loss of unlimited proliferative ability' and despite the spectrum of 'deaths' that irradiated cells may die, survival curves for all types of mammalian cells *in vitro* are remarkably similar.

Figure 3.4 shows a generalized survival curve for radiation-induced loss of reproductive capacity (cell death) in mammalian cells. The typical survival curve for mammalian cells (figure 3.4*b*) has a 'shoulder' at low doses and only becomes exponential at higher doses. The D_0 determines the slope of the linear part of the curve and indicates the dose that will reduce cell survival by a factor of 0·37. The extrapolation number (N) is derived by extrapolating the straight-line portion of the curve on to the survival axis and N is seen to be 3 in figure 3.4*b*. The majority of mammalian cell survival curves, following low LET radiation, have D_0 values of between 1 and 2 Gy and extrapolation numbers of between about 1 and 5. However, the size of the shoulder and the slope of the curve can vary even for a given cell line depending on growth conditions.

Since the shape of a survival curve is fully defined by the two terms D_0 and N they are the most useful parameters when comparing different curves, for example between normal and tumour cells. A third parameter is D_q—the quasi-threshold dose. D_q is the dose in gray where the exponential part of the curve crosses the 100 per cent survival (see figure 3.4). It is a measure of the size of the shoulder and as we shall see in the next chapter its importance is related to the repair of radiation damage. Most D_q values are between 0·5 and 2·5 Gy for acute X-irradiated, well oxygenated cells.

3.5. Target theory and mammalian survival curves

The simple exponential survival curve for viral inactivation was described using target theory as a 'single hit curve'. However, mammalian cell survival curves are more complex and generally fit what are called multi-event models. Unfortunately the application of target theory to mammalian cell survival curves is limited by the statistical errors associated with the determination of the points of the curves (see figure 3.5). The errors are particularly significant at low doses. The

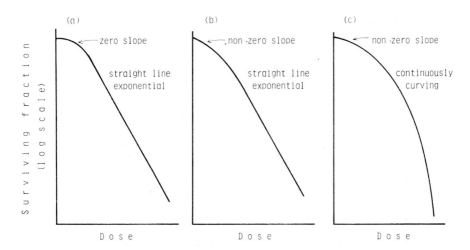

Figure 3.5. Three of the models used to fit experimental mammalian cell survival data. (*a*) Multi-target; (*b*) Multi-target plus single hit component; (*c*) Continuously curving curve

uncertainties that cause the errors include the randomness of cell killing which in turn is determined by 'microdosimetric' considerations that involve statistical variations in energy deposition in such small volumes as cell nuclei. There are also biological factors including variations in the intrinsic radiosensitivity of the phase of the cell cycle (see Chapter 4) and sampling and cell dilution errors that are inherent in the survival assay techniques. So the application of target theory is not straightforward. Despite this, the study of the shape of survival curves is an essential object of quantitative radiobiology. In addition, it has direct practical relevance to radiation protection problems of the hazards of low dose (see Chapter 12) and is of importance to radiotherapy where doses of 2–3 gray are given daily over a 4–6 weeks period (see Chapter 9).

The major effort is directed towards defining the initial shape of the killing curve at low doses. Figure 3.5 illustrates three of the possible mathematical models used in the analysis of survival curve data.

The simple multitarget model

The first *in vitro* survival curve produced in 1956 showed an excellent fit to a multitarget model. This model suggests that two or more targets must each receive a single hit before the cell is killed and can be described by the algebraic equation

$$f = 1 - (1 - e^{D/D_0})^N$$

where f is the surviving fraction after a dose D, D_0 is the 37 per cent dose slope and N the extrapolation number. N is also formally equivalent to the number of targets in a cell that have each to be inactivated by a single hit. The model implies a zero slope to the curve at zero dose which means that the lowest dose will produce no cell killing. However, the majority of published experimental data do not support an initial zero slope.

The multitarget, single hit model

This model fits the majority of the mammalian cell survival curves and implies the possibility of single and multi-hit inactivation

$$f = e^{-D/D_1}[1 - (1 - e^{-D/D_2})]^N$$

Its significant feature is that it indicates a non-zero initial slope at low doses preceding the straight line exponential, i.e. there is cell killing even at the lowest doses. This model suggests that there are a variety of processes that can kill cells. It combines the principles of the simple multitarget model with a single event component and it is the latter that gives the non-zero slope at low doses. This single hit component is usually attributed to the action of the high LET radiations that are part of the LET spectra of the charged particles associated with the absorption of X- and γ rays and fast electrons (see Chapter 1). It is postulated that these high LET events may have enough energy to inactivate all targets in a cell at once or cause the several hits on a single target, if that is what cell killing involves. Probably a majority of mammalian cell survival curves can be fitted by the equation above.

The continuously curving curve

This model is represented by the linear quadratic equation

$$f = e^{-(\alpha D + \beta D^2)}$$

where α and β determine the relative importance of inactivation from a single hit and from two hit events. The model indicates that there is a negative initial slope preceding a continuously curving curve at higher doses. The extent of the curviness being a function of the relative values of α and β. There are relatively few published survival curves that truly conform to this model over a wide range of doses.

The statistical considerations mentioned at the beginning of this section do not allow us to decide at present which, if any, of these models is most likely to be correct. The majority of experiments support the multitarget, single hit model and the idea of a non-zero slope for the shoulder region at the lowest doses. The main challenge to the multitarget, single hit model comes from the linear quadratic model. The latter derives its support from the so-called 'dual theory of radiation' whose microdosimetric hypotheses are beyond the scope of this book. However, critical experiments have recently cast doubt on these hypotheses and the whole field is currently in a state of flux. Indeed, in a few years time it may well be true, as was recently stated, that "all survival curves are basically one hit exponentials and the shoulders at low doses and the variability of shapes are both due to various rates, efficiencies and types of repair processes" (see Chapter 4).

3.6. Is DNA the target for cell killing?

As we noted above, experimental data are often too imprecise to allow one to choose between the various target models and this is particularly true of doses in the 0·5–1·0 Gy dose level. However, the consensus of current opinion would probably favour the last two models since they both involve a non-zero initial slope. We shall be returning to the importance of threshold values and low dose effects in Chapters 8 and 11.

Despite the vagueness of the terms 'hit' and 'target' when applied to mammalian cells, target theory remains a useful way of looking at cell survival. Due to the prime importance of DNA to the cell, it is hardly surprising that DNA should be regarded as the most likely target for cell killing; and there is a mass of evidence to support such a claim.

(a) There is little doubt that the inactivation of viruses and bacteriophages (bacterial viruses) is a result of damage to their DNA. Research with bacteriophages has shown that double strand breaks in DNA are most likely to have lethal consequences. There is also evidence that unrepaired single strand breaks and some types of base damage can be lethal. In bacteria the evidence for DNA as the target is somewhat equivocal and some research workers incline to the view that the bacterial cell membranes may be the primary site of radiation damage.

(b) In more complex cells than micro-organisms the identity of the target is a source of some argument. There is, however, a certain amount of circumstantial evidence that points to DNA as the critical target for cell lethality. For example, evidence was cited in the last chapter (p. 47) that showed that the nuclear material is far more radiosensitive than the cytoplasmic material in relation to cell killing.

(c) Data were also given in the last chapter on the cell killing effect of the incorporation of ^{3}H- and ^{125}I-labelled compounds into the DNA molecules. This implicates chromosomal material because most of the energy of the electrons from these isotopes is dissipated in the chromosomes.

(d) There is a mass of data for fungi, insects and higher plants and animals that suggests a connection between cell killing and ploidy. A change in ploidy very often causes a modification in the radiosensitivity for cell killing. This implicates DNA since ploidy is merely the word used to indicate the relative chromosome content of cells.

(e) Another strong piece of evidence that implicates DNA as the target for cell killing comes from the use of halogenated pyrimidines. These are pyrimidines in which a halogen has been substituted for the methyl group and the most commonly used is 5-bromodeoxyuridine (BUdR). Since BUdR is almost identical with the thymine molecule it is specifically incorporated into the structure of DNA. However, because BUdR is *not* identical with thymine, its incorporation sets up stresses in the DNA that increase the yield of strand breaks. As more and more BUdR is incorporated into the DNA, the cells become increasingly radiosensitive (see figure 8.10, p.145).

(f) The most durable hypothesis attributes radiation-induced cell killing to damage to the genetic apparatus, specifically to chromosome aberrations, and chromosome damage is the final expression of some forms of DNA damage (for details see Chapter 7, section 4). There is a close, almost 1:1 correlation between cell death at the first mitosis after radiation and the presence of chromosome aberrations. This 1:1 relationship is maintained under a variety of modifying conditions, for example, phases of the cell cycle, in the presence and absence of sensitizing agents and so on.

(g) There is increasing evidence in a number of human diseases of a correlation between defective or altered DNA repair mechanisms and the enhanced sensitivity to radiation of both the patients and their isolated cells. Figure 3.6 shows the extreme sensitivity of fibroblasts from persons with Ataxia telangiectasia compared with cells from normal persons (see p. 45).

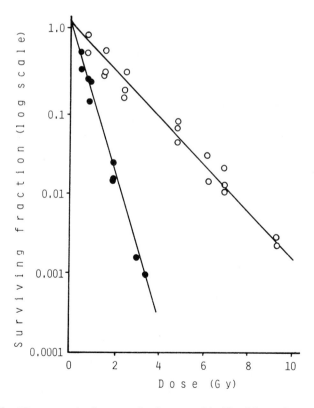

Figure 3.6. X-ray survival curves for human skin fibroblasts from normal patients (○) and from patients with Ataxia telangectasia (●)
Source: R. R. Weichselbaum, J. Nove and J. B. Little, 1977, *Nature* **266**, 726; courtesy the authors and publisher.

We may conclude that there is much evidence that points to DNA as the primary target for cell killing. However, there are few radiobiologists who would agree that DNA is the only target and many would wish to include the nuclear membranes as critical sites of damage.

In Chapter 2 we discussed the major types of radiation-induced DNA damage and repair mechanisms and noted that after doses of about 1–2 Gy, i.e. D_0 doses for cell killing, there would be an "excess of DNA damage"—about 1000 single strand breaks and perhaps 50 double strand breaks. It has been suggested that surviving cells repair most if not all of these breaks and that cell death is due to lack of repair or misrepair of a small fraction of breaks.

At present there is no agreement over what type of DNA damage is most relevant to cell death. One theory assumes that the critical lethal lesion is the double strand break (DSB) that may arise either by a single hit or by two hits occurring opposite one another in the complementary strands of the DNA (see Chapter 2). This theory implies that the α and β terms in the linear quadratic equation

$$f = e^{-(\alpha D + \beta D^2)}$$

describe the induction of single strand and double strand breaks. At its simplest this molecular theory of cell killing states that DSBs cause chromosome aberrations and that aberrations cause cell death. The theory essentially hinges on the mathematical similarity between the linear quadratic relationships that have been obtained at least in certain studies and under certain conditions for:

(a) DNA double strand breaks (see Chapter 2).
(b) Chromosome aberrations (see Chapter 7).
(c) Cell killing—see the equation above.

However, many workers disagree with many of the assumptions of the theory, such as "DSBs are the most important radiation lesions in DNA", "DSBs are the chief cause of chromosome aberrations" or "reproductive death of cells is due solely to radiation-induced chromosome abnormalities".

So the role of DNA strand breaks as the cause of cell death remains controversial. There are studies that show a close correlation between DNA breaks and cell death whereas other studies find no such correlation. Many studies are equivocal because the very high doses needed for strand break analysis are far in excess of those needed for substantial cell killing. Recent advances that allow breaks to be detected at moderate doses ($\geqslant 0.5$ Gy) are producing important insights into break production and repair. One such study found that in two cell lines with very different sensitivities for cell killing there were no such differences in sensitivity for either the initial or residual (i.e. after repair) number of

DNA breaks. The controversy about cell viability and DNA double strand breaks rests upon the assumption that it is difficult to conceive of a viable cell with its DNA in pieces. However, many cells do survive with some chromosomes fragmented or even with whole chromosomes missing. Nevertheless, the molecular theory of radiobiology has been expanded to cover virtually the whole of radiation effects. The theory suggests that double strand breaks in the DNA are responsible for chromosome abnormalities and genetic mutations (see Chapter 7), for cell death and for the induction of cancer by radiation (see Chapter 9). Many workers remain unconvinced by the universality of the hypothesis and feel it is altogether too simplistic.

3.7. Summary

Viruses, bacteria, plant and animal cells show a loss of reproductive ability following radiation. The dose needed to inactivate ('kill') 63 per cent of a population of cells, the D_0, ranges from 500 Gy for viruses down to 1-2 Gy for mammalian cells.

The criterion for survival in the case of mammalian cells *in vitro* is the ability of the cell to produce, by division, a visible macroscopic colony of some 32-64 cells. Cells failing to do this are called 'radiation-killed cells'. It must be remembered that such cells need not be metabolically inert. The 'dead' cells suffer a variety of fates, all of which effectively inhibit rapid colony formation and condemn the cell to be scored as dead.

Radiation energy is deposited in living matter as random ionization and excitation events. Target theory states that cells contain one or more critical sites or 'targets' within which an ionization event would be fatal to the cell. Ionization events outside the target do not cause cell death; target theory strictly applies to the direct action of radiation. The diffusion of free radicals as the mediators of radiation damage and the ability of the cell to repair the initial radiation damage make the simple application of target theory to higher plants and animal cells difficult. Nevertheless, mammalian cells do comply quite well with some target models and so the search for a target molecule has not abated, and much evidence points to DNA. The intimate association between DNA and nuclear membranes also highlights the probable crucial role of membranes in cell killing.

Chapter 4
Cellular effects of radiation (II)

4.1. The radiosensitivity of the different phases of the cell cycle

The *in vitro* methods of measuring the cell-killing effect of radiation were described in the last chapter. In the Puck method for mammalian cells a population of single cells is plated onto a petri dish. Following different doses of radiation each cell's ability to divide and produce macroscopic colonies is determined. If a cell forms such a colony it is said to have survived; if it fails to form a colony it is said to have been killed by radiation. Normally the cells cultured for such experiments are distributed randomly around the cell cycle, the stages of which—M, G_1, S and G_2—were discussed in Chapter 2.

In contrast to such randomly dividing cell populations, it is possible to obtain populations in which all the cells divide synchronously and move around the cell cycle at much the same speed. In this chapter we shall describe experiments that have been performed to test the relative sensitivity of populations of mammalian cells in which all the cells are, for instance, in mitosis, or all in DNA synthesis, and so on. So far this work has only been feasible *in vitro* since most cell populations *in vivo* divide asynchronously. There are only a few instances of synchronous cell division *in vivo*, for example, the first few divisions of the developing fertilized ovum are synchronous up to about the 64-cell stage, after which divisions become somewhat irregular, and following massive injury to the liver, the regenerative response of the parenchymal cells involves one or two waves of relatively synchronous cell divisions. Populations of cells *in vitro* and *in vivo* tend to be asynchronous because each cell varies just a little from its neighbours in the speed with which it progresses through the cycle.

66

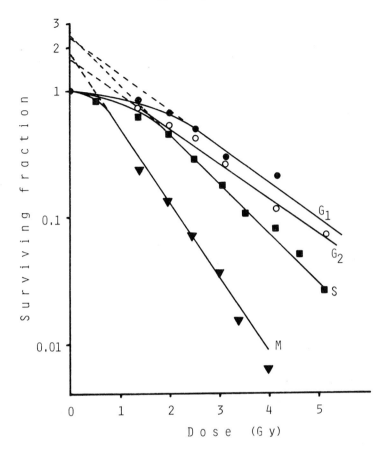

Figure 4.1. **Survival of the reproductive capacity of X-irradiated HeLa cells at four different times during the cell cycle; M (mitosis), G_1, G_2 and DNA S phase**
Source: T. T. Terasima and L. J. Tolmach, 1963, *Biophysical Journal* 3, 11; courtesy the authors and the Rockefeller Institute Press.

4.2. The radiosensitivity of synchronized cell populations

There are a number of ways *in vitro* to collect cells into one particular phase of the cell cycle and to follow them through the cycle as they move synchronously through the phases. Strict synchrony lasts only for one or two cell generations (one or two division cycles) and then the individual differences between cells lead to its gradual loss.

The first experiment to test the effect of the phase of the cell cycle was reported in 1963 with a human cancer cell line known as HeLa cells. In the experiment a single dose of 3 Gy was given to populations of cells in specific phases of the cycle. This immediately showed that the killing of cells by radiation depended markedly on the phase at which they received the radiation. Cells in mitosis were the most sensitive, and cells in early G_1, late S and G_2 were the most resistant. Cells at the beginning of S were of intermediate sensitivity. The work was confirmed using a series of different doses, and producing complete survival curves for the specific phases. Figure 4.1 gives the results of these experiments. The series of curves shows how the radiosensitivity of the cells changes relative to the phase in which they were at the time of radiation. The D_0 of HeLa cells in M phase is 0·82 Gy, for cells in G_1 it is 1·64 Gy, for cells in mid S phase it is 1·21 Gy and for cells in G_2 it is 1·7 Gy.

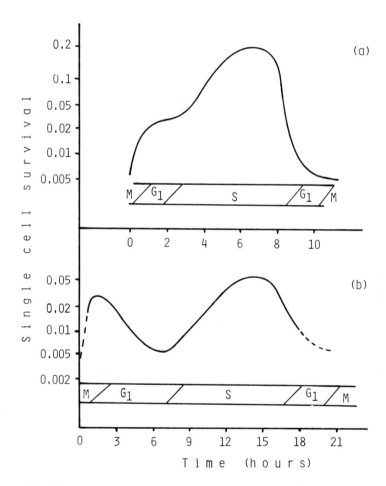

Figure 4.2. The relative radiosensitivity of the phases of the cell cycle for (a) cells with a short G_1, represented by Chinese hamster cells; and (b) cells with a long G_1, represented by HeLa cells

Source: W. K. Sinclair, 1969, in *Comparative Cellular and Species Radiosensitivity*, edited by Bond *et al.* Igaku Shoin Ltd., p.69; courtesy the author and publisher.

Since this initial study other workers have used other cell lines and other methods of producing synchrony in their cells. Not surprisingly each cell line seems somewhat specific with respect to the relative radiosensitivity of the different phases of the cycle. Figure 4.2 shows the forms of phase response in two types of cells: (*a*) those having a short G_1 represented by a Chinese hamster line after 7 Gy; and (*b*) those having a long G_1 period represented by HeLa cells after 5 Gy.

Generally speaking there are differences of a factor of two in the dose required for a given level of cell killing between the different phases of the cycle. Cells with a short G_1 are usually least sensitive in the latter part of S phase, more sensitive in G_1 and most sensitive in mitosis and G_2. This pattern of relative sensitivities also holds for cells with a long G_1 except that there is also a resistant phase in early G_1 and a sensitive stage at its end.

It has been shown that following exposure to high LET radiation, for example, fission neutrons and α particles, the variation in radiosensitivity of the different phases of the cell cycle is much less than for low LET radiation.

4.3. Repair of cells following radiation

There are at least three types of repair which increase the post-irradiation survival of cells. They are: repair of sublethal damage (SLD), the repair of potentially lethal damage (PLD), and 'slow repair'.

Repair of sublethal damage

In discussing the theoretical aspects of mammalian cell survival curves in Chapter 3 we saw that they were characterized by a shoulder at the low-dose region followed by an exponential portion at higher dose levels. In terms of target theory this shape suggests that the killing of the cells requires one or more hits on each of several critical targets; the shoulder indicating that the fraction of cells killed per unit dose is much less in the initial shoulder region than at higher doses.

At very low doses, although there is a finite probability that some cells will be killed, most suffer what is called sublethal damage, which in target theory terms means they have not received a sufficient number of hits to kill them. However, an alternative hypothesis which is becoming increasingly attractive attributes the shoulder to a repair process known as 'Q repair', which becomes less and less effective as the dose increases. Cells lacking Q repair will be killed exponentially even at the lowest doses. If such repair ideas are accepted then all survival curves could be one hit exponentials and the shoulders at low doses represent the possibility of the repair of sublethal damage. Such a repair model would also imply that N, the extrapolation number, has nothing to do with hits or targets but might merely reflect the concentration of Q factor, for example, a pool of repair enzymes or of radical scavenging molecules. The latter might be used up or 'saturated' by the first few gray of radiation and exponential killing would follow at higher doses.

Figure 4.3 shows a mammalian cell survival curve following a series of single doses of radiation—a dose of 10 Gy is shown as reducing survival to level S. If instead of giving a single dose of 10 Gy a smaller dose of 5 Gy is given the level of survival S_1 will be produced. If a few hours elapse before a second dose of 5 Gy is given, the level of survival will be S_2. In both cases the cells have received a total dose of 10 Gy either as a single dose or as a split (or 'fractionated') dose of 2×5 Gy. The level of survival is much higher after the split doses than after the single dose. The

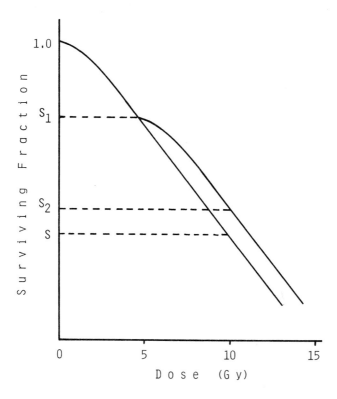

Figure 4.3. A survival curve following a single dose of 10 Gy reducing survival to level S, and the effect of splitting the 10 Gy into two equal fractions separated by some time interval such that S_1 and S_2 are the levels of survival produced by the first and second dose of 5 Gy respectively

explanation for this is that the cells are able to recover from a certain fraction of the damage produced by the first dose of radiation before the second exposure. The recovery of the cells is indicated by the reappearance of the shoulder of the survival curve between the first and the second doses. If no recovery had occurred there would be no decrease in effect by splitting the dose. The first dose of 5 Gy would of course have reduced the level of survival to S_1; but the second 5 Gy would have led to a survival level S identical to that produced by a single dose of 10 Gy.

Mammalian cells are able to recover from some of the sublethal damage produced by the first dose so that a second dose is received by a more or less intact, undamaged population of cells. Since these cells are now almost undamaged, a second dose fraction of radiation will for many cells only be sublethal in effect.

In the case of survival curves with little or no shoulder, such as those obtained following high LET radiation, little or no recovery is obtained by splitting the dose. This is because the high LET radiations are so

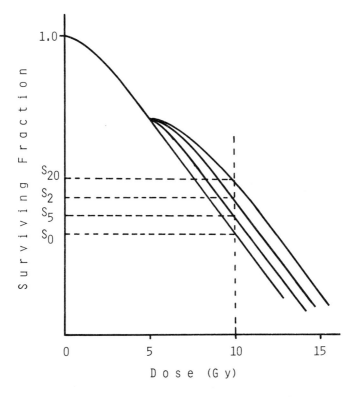

Figure 4.4. Survival curve showing the effect of the time interval between the first and the second dose of 5 Gy. S_0, S_5, S_2 and S_{20} represent the level of survival for time intervals of 0, 5, 2 and 20 hours respectively

efficient at killing cells, however small the dose (see Chapter 8).

For low LET radiation the time interval between the first and the second dose is of great importance in determining the amount of recovery of the cells, and this is seen in figure 4.4. This shows the single dose curve for doses up to 10 Gy and the effect of giving a first dose of 5 Gy followed at either 0 h, 2 h, 5 h or 20 h intervals by a series of other doses up to a total dose of 10 Gy. The 0 h curve is equivalent of course to the single dose curve and the 20 h curve is displaced to the right by the almost total reappearance of the shoulder. It should be noted that the reappearance of the shoulder is not accompanied by any change in the slope, i.e. the D_0. The survival levels S_0, S_2, S_5 and S_{20} are marked off on figure 4.4 and show that the reappearance of the shoulder, i.e. the recovery from sublethal damage, is a time-dependent phenomenon.

Figure 4.5 plots the relative surviving fraction against the time interval between the two doses. If the time interval is zero hours, i.e. a single dose, the proportion of cells killed is greater than for any other

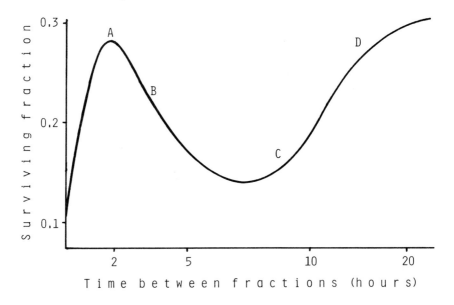

Figure 4.5.   The fluctuating recovery curve of cells as a function of the time interval between fractionated doses

A=phase of intracellular recovery.

B=the phase when partially synchronized and predominantly resistant population of cells begins to move around the cell cycle again, becoming more and more sensitive to the second half of the dose.

C=the phase where the semi-synchronous population becomes more resistant as it moves around the cycle.

D=the final phase represents the time when the cells are losing all semblance of synchrony and are beginning to multiply.

time interval. As the interval increases an increasing proportion of the cells survive until at 2 h it is a maximum. This indicates that the repair of sublethal damage is rapid and suggests that it may not be dependent on the cell cycle. This suggestion is borne out by experiments designed to inhibit the activity of mammalian cells by keeping them at room temperature (24°C). The results of a split dose experiment at 24°C are given in figure 4.6, which indicates that the initial rapid rise of the recovery curve—the phase of intracellular repair—is much less affected by temperature than are the subsequent fluctuations. The possible mechanisms involved in the intracellular repair of sublethal damage have been extensively studied.

Experiments carried out with chemical inhibitors of such cellular processes as DNA, RNA and protein synthesis led initially to the conclusion that the repair process was non-enzymatic—that it was a purely physico-chemical process which did not require metabolic intervention. However, recent studies with metabolic inhibitors and more particularly with oxygen have led to the opposite conclusion; namely that repair of sublethal damage does involve cellular metabolism.

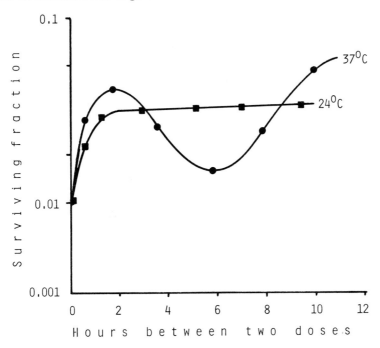

Figure 4.6.   The recovery curve for cells kept at room temperature (24°C) and at 37°C
Source: M. M. Elkind, 1966, in *Radiation Research*, edited by G. Silini, North Holland Publishing Co.; courtesy the author and publisher.

In a number of rigorously controlled experiments it has been shown that the presence or the virtual absence of oxygen makes no difference to the curving part of the survival curve. This implies that extremely hypoxic cells do accumulate sublethal damage. However, repair of this damage is very much reduced or even absent if the cells are kept free of oxygen in the interval between the dose fractions. Presumably the lack of oxygen causes a reduction in aerobic respiration, which leads to an insufficient supply of energy for the metabolic steps involved in the repair of sublethal damage. Repair is not affected by the presence or absence of oxygen at the time of irradiation. So, repair of sublethal damage is oxygen-dependent, is reduced at temperatures below 37°C (see figure 4.6) and can be inhibited by certain metabolic blocking agents, all of which suggests that it does involve cellular metabolism.

The molecular events responsible for sublethal lesions and their rapid repair are still a matter of conjecture. The rapid repair phase A in figure 4.5 has just been described and the other phases, B, C and D, must now be dealt with briefly.

In order to account for the fluctuations B, C and D we have to go back to what was said about the relative radiosensitivity of cells in different phases of the cell cycle (p. 68).

The first dose of radiation will preferentially kill the most sensitive cells, leaving a population of cells predominantly in one or two resistant parts of the cell cycle. Radiation therefore will partially synchronize the cells, but it will also inhibit movement around the cell cycle for a short while, say 2–3 hours, depending on the dose. When this population of resistant cells begins to move around the cycle the cells will all tend to move to more sensitive stages of the cycle. It is this population getting more and more sensitive that will be exposed to the second dose of radiation at some 3, 4 or 5 hours after the first dose. The result is that phase B of figure 4.5 is produced.

Eventually, the semi-synchronous population moves to a more resistant stage, resulting in the rising phase C. Finally, the cells become more or less desynchronized and their response to a second dose of radiation is a summation of the individual phase sensitivities: final phase, D.

This fluctuating recovery pattern is common to most mammalian cells, only the details of the timing of the peaks and troughs of the curve varying from cell type to cell type.

The repair of sublethal damage after high LET radiation is much reduced but follows the same time course. This lack of repair is commensurate with the smallness of the shoulder on high LET survival curves (see Chapter 8).

Finally, several experiments have been done using repeated fractions, i.e. mimicking clinical radiotherapy practice (see Chapter 9). Such experiments are designed to discover whether cells are capable of full recovery time and time again. Some experiments suggest they are, while others suggest that there is a gradual fall in recovery capacity after say five or more fractions.

One important practical aspect of the knowledge of the ability of cells to recover from sublethal doses of radiation is its application to the radiotherapy of cancer.

One interesting speculation is that tumours are cured in radiotherapy because the multiple dose regimes eventually reduce the tumour's capacity to recover (see also Chapter 9).

Repair of potentially lethal damage

The repair of potentially lethal damage (PLD) can be demonstrated both *in vitro* and *in vivo* using single doses of radiation and then altering the post-irradiation conditions. The conditions which favour PLD repair are those which are 'suboptimal', i.e. which merely allow cells to tick over but do not encourage cell growth or cell division. So, post-irradiation conditions such as low temperatures, metabolic inhibitory drugs or putting cells in saline after radiation instead of the usual nutritive growth media will all allow a greater fraction of cells to survive than if the cells are maintained under optimum conditions.

PLD repair can be seen best *in vitro* with so-called plateau phase cells rather than with normal rapidly growing *in vitro* cultures. When cells are grown in monolayers in culture they initially multiply exponentially and there is little cell loss or death. However, if the nutrient medium is not replaced, growth slows and the cells gradually reach a saturation cell density. The culture is now in a 'stationary' or 'plateau' phase in which growth is severely limited, the majority of the cells being either arrested in G_1 or progressing very slowly in other phases. Using such cultures it has been shown that the alteration of post-irradiation conditions can reduce the amount of radiation cell killing. The experiments involved irradiating plateau-phase cells with single doses and then sub-culturing them onto petri dishes and scoring them about ten days later for clonal survival (see Chapter 3). If, immediately after radiation, there is a delay of a few hours before sub-culturing and assaying the cells for survival, then the level of survival of the cells is enhanced.

Figure 4.7 shows the surviving fraction for a mammalian tumour cell line irradiated with graded doses of X-rays or neutrons and sub-cultured 'immediately' or after a 'delay' of 5 hours. The X-ray data show clearly that PLD repair has occurred while for neutrons PLD repair is almost absent. It should be noted that the repair of potentially lethal damage is accompanied by a change in D_0 and in this respect it differs from repair of sublethal damage in which no change of D_0 is observed. The virtual absence of PLD repair after high LET radiation is characteristic for most, but not all, cell lines.

The X-ray curves in figure 4.7 also illustrate the fact that repair of PLD involves an increase in D_0 with little or no change in the shoulder region. The molecular nature of potentially lethal lesions is unknown. The lesions could become lethal by, say, misrepair of DNA or enzymes released from damaged membranes, or non-lethal by, for example, correct repair of

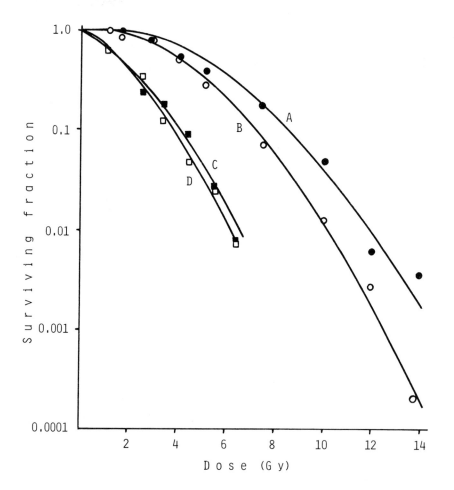

Figure 4.7. Survival curves for cells given X-rays (curves A and B) and neutrons (curves C and D)
Source: J. S. Rasey and N. J. Nelson, 1981, *Radiation Research* **85**, 69; courtesy the authors and Academic Press.
For Curves A and C there was a delay of 5 hours between irradiation and the subculture of the cells for colony assay. This delay allowed recovery from potentially lethal damage to occur for X-rayed cells (A) but not for neutron irradiated cells (C).

DNA or membrane damage. In this context it is interesting to note that cells taken from patients with Ataxia telangiectasia are deficient in the repair of PLD after X-irradiation (see also Chapter 3).

Slow repair

In radiotherapy it has also been found that less damage is produced in normal and tumour tissues by extending the treatment time over a period of several weeks. It used to be thought that such 'sparing' effects were due to repopulation of the tissues by surviving cells. However, recent work suggests that a process of 'slow repair' may also operate which is similar to, but about 100 times slower than, the rapid repair of sublethal damage described above. Such slow repair occurs in the lungs and kidney but probably not in the spinal cord or bladder. Other slow types of tissue repair have been reported in bone and blood vessels. The mechanisms of these slow repair phenomena remain obscure.

4.4. Summary

In this chapter, two important responses of cells to radiation have been discussed.

First, it is clear that the sensitivity of cells to radiation killing depends on the stage of the cell cycle in which they are at the time of radiation. Different cell lines vary in the relative radiosensitivity of the phases of their cell cycles.

Second, under certain conditions, cells are capable of three types of repair which increase their post-irradiation survival chances. These are: the repair of sublethal damage (SLD), the repair of potentially lethal damage (PLD), and certain forms of 'slow repair'.

Repair of SLD is demonstrated using split doses—the reduced effectiveness of the latter shows that cells are capable of repairing a certain fraction of damage. The shoulder region of the survival curve suggests the accumulation of sublethal damage and the possibility of the repair of such lesions. The time between the first and second parts of the dose produces a characteristic fluctuating curve for the recovery after low LET radiations. The molecular mechanisms (DNA/membrane repair?) responsible for SLD repair remain unknown but they are dependent upon oxygen availability and active cell metabolism. The trough and subsequent rise of the recovery curve are thought to be due to changes in the phasic sensitivity of cells.

PLD repair is demonstrated using single doses, but so altering the post-irradiation conditions of the cells that a greater fraction of cells survive than if the cells were maintained under optimum conditions.

Both SLD and PLD repair are generally reduced in cells irradiated with high LET radiation, and both repair mechanisms have been shown to occur *in vivo* as well as *in vitro* for normal and tumour cells.

Chapter 5
Radiation cell survival *in vivo*

5.1. *Cells* in vitro *and cells* in vivo

A great deal of fundamental radiobiology has been done using cells cultured *in vitro* and some of this work was described in Chapters 3 and 4. But are these *in vitro* results applicable to cells *in vivo*? Do the cell survival data obtained with isolated cells *in vitro* bear any relationship to cell killing *in vivo*? These questions will be considered in this chapter, but first let us look at some of the reasons for supposing that cells that have become established *in vitro* differ from tissues *in vivo*.

First, the established cell lines *in vitro* have been selected for their ability to grow in the culture conditions and optimum growth is observed at between 5×10^4 and 5×10^5 cells cm^{-3} of liquid medium. This compares with the much higher cell densities of 10^8 cells cm^{-3} found in most solid tissues. Cells *in vivo* grow in an integrated way, their closeness to one another allowing a variety of interactions between neighbouring cells, while cells *in vitro* generally grow in isolation from each other and there is little feedback between one cell and another.

Second, cells *in vitro* proliferate on average much more rapidly than cells *in vivo*. *In vivo* frequent cell division is only observed in certain cells of, for example, the bone marrow, intestine and skin. In most organs of the adult body, cell divisions are infrequent and there may be weeks, months or even years between one division and the next.

Third, besides these differences it is known that in certain cases normal and especially tumour tissues do not grow in such uniform conditions as those which prevail in cell culture. *In vitro* cells are in an environment where there is an ample supply of oxygen and nutrients, where the temperature, humidity and pH are at optimal levels and where there is space for unlimited proliferation. *In vivo* the blood supply is not necessarily uniform (again, especially in tumours) and consequently nutritive and gaseous supplies may be irregular.

Finally, cells *in vivo* are subject to immunological and hormonal factors that are not present in culture conditions.

These differences mean that it is not possible simply to say that since cells are killed *in vitro* by radiation in such a manner the same effects will occur *in vivo*, and this fact led radiobiologists to develop *in vivo* methods of assessing cell survival.

5.2. Three-dimensional 'spheroid' cultures

Before describing a number of *in vivo* assays of cell survival we should mention a potentially important method that is halfway between the *in vitro* and the *in vivo* methods. If cultured cells are held in suspension by being stirred they may aggregate into what are called spheroids. These are three-dimensional clusters of hundreds or thousands of cells that have many of the characteristics of solid tumours and are therefore of great value in studying aspects of the radiation therapy of cancer (see Chapter 9). For example, cells in spheroids show a wide distribution of cell cycle times, larger spheroids have fewer and fewer cells in active proliferation, there may be areas of dead or dying cells and there may be areas deficient in oxygen (hypoxia). The technique of the assay involves irradiating the spheroids, disaggregating them and then assaying the individual cells by the standard *in vitro* technique described in Chapter 3. Cells grown as spheroids are generally more resistant to radiation than single cells in culture. This radioresistance seems to be related to an enhanced ability to sustain sublethal damage—the shoulder of the survival curve of spheroid grown cells is larger ($D_q = 4$–9 Gy) than the shoulder region of cells in monolayer culture ($D_q < 2$ Gy). This shoulder width is in general agreement with survival curves of organized tissues which have D_qs of 4–6 Gy. The enhanced repair capacity of semi-organized tissues *in vitro* (spheroids) and organized tissues *in vivo* may be related to subtle cell-to-cell contacts that are disrupted in single-cell cultures *in vitro*.

Some workers have reported not only a broadening of the shoulder of the survival curve of spheroid cells but also an increase in the slope (D_0) of the curves.

Figure 5.1 shows two survival curves, one for cells from spheroids 300 μm in diameter and the other for cells from spheroids 550 μm in diameter. The former is purely exponential after the shoulder region while the curve for cells from the larger spheroid has a 'resistant tail'. This tail represents the presence of a small fraction of radioresistant hypoxic cells in the larger spheroids; the significance of the oxygen effect and of such cells will be discussed in Chapters 8 and 9. A recent review article on spheroids has stressed the need for caution in deducing general principles from

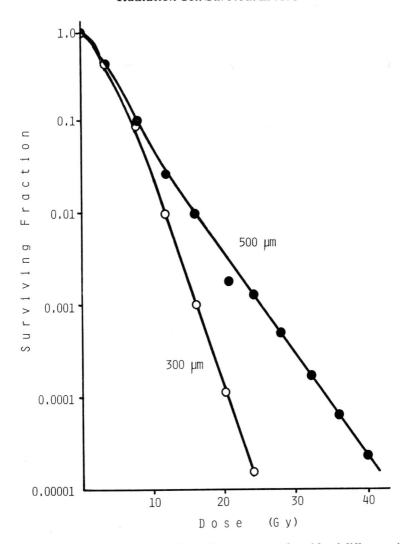

Figure 5.1.   X-ray survival curves for cells grown as spheroids of different sizes
Source: R.E. Durand, 1981, *Radiation Research* **1**, 85; courtesy the author and Academic Press.

spheroid model systems and shows how dramatic differences in radiation response can be produced by minor changes in technique.

5.3. The 'tumour dose-50' method

The first method for the quantitative determination of the reproductive capacity of tumour cells *in vivo* was devised by using a certain type of mouse leukaemia cell which grows as a single cell suspension and which infiltrates the liver. A 'donor' mouse with advanced leukaemia is killed and its liver removed. The liver is minced up and varying numbers of leukaemia cells are then intravenously injected into genetically similar (syngeneic) 'recipient' mice and if enough cells are used the recipients will develop leukaemia as a result of the proliferation of the injected cells. The recipients are given up to 90 days from the time of the injection in which to show signs of leukaemia, and a mouse developing leukaemia is called a 'take'. By varying the number of leukaemia cells it is possible, using certain statistical methods, to calculate how many cells it is necessary to inject into a group of mice in order to get 50 per cent 'takes'. This number is known as the 'tumour dose-50' or TD_{50}, and in the case of the leukaemia cells above the TD_{50} is 2 cells. This means that if 100 mice are each injected intravenously with 2 leukaemia cells, then on average 50 of them will develop leukaemia. Using this figure of 2 it is possible to build up a radiation survival curve. Following whole body irradiation of the leukaemic donor mouse it is killed and the leukaemia cells are extracted, counted and varying numbers of them are injected into groups of recipient mice. These recipients are then observed as previously for signs of leukaemia over a 90-day period. If the radiation prevents cell division in a proportion of the leukaemia cells then proportionately more cells must be injected into the recipients in order to produce the same percentage 'takes' as was the case for the mice receiving unirradiated cells. If 2 cells is the TD_{50} for cells given no irradiation, and if it is found that following a dose of radiation 20 cells have to be injected to give 50 per cent takes, then it would appear that 18 out of 20 cells have been inactivated by the radiation, i.e. 90 per cent radiation inactivation, 10 per cent survival. Figure 5.2 shows a survival curve for leukaemia cells using the TD_{50} method. It has a D_0 of 1·6 Gy and is exponential down to very low levels of survival.

The D_0 of 1·6 Gy for this leukaemia cell line determined by the TD_{50} method *in vivo* is strictly comparable with the D_0s obtained for mammalian cells using the *in vitro* technique described in Chapter 3. The criterion of survival in the Puck plating method is whether a cell completes 5 or 6 successful post-irradiation divisions producing some 32 to 64 progeny. The criterion of cell survival in this *in vivo* TD_{50} method is whether the cell can divide and produce the several millions of progeny that result in an animal being classed as 'leukaemic'. However, the two criteria are similar since it is known that cells capable of 5 or 6 divisions *in vitro* are capable of almost unlimited proliferation. In the case of solid tumour cells, the TD_{50} criterion is the formation of a solid tumour in recipients that have received cells from a solid tumour. The method has been used over the past 20 years in a great number of different tumours. A comparison of the survival curve parameters (D_0, D_q and N) shows

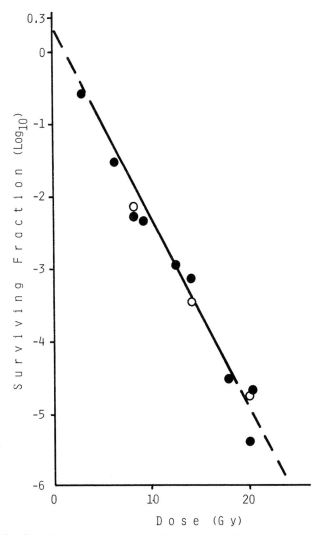

Figure 5.2.   Survival curve for mouse leukaemia cells irradiated *in vivo* **with cobalt-60** γ **rays**
Source: H. B. Hewitt, 1962, *Scientific Basis of Medicine: Annual Review* 1, 303; courtesy the author and Athlone Press.

there are really no differences between tumour cells assayed either by the *in vitro* or the *in vivo* TD_{50} method.

5.4. The spleen nodule method

The TD_{50} method is only applicable to malignant tumour cells since it tests the ability of a cell or group of cells to produce a tumour in a given time interval. It is possible, however, to measure the radiation response of normal cells *in vivo* and one of the best-known methods is the spleen nodule technique for bone marrow cells.

In mammalian bone marrow there are a certain number of cells whose role it is to replace the different types of blood cells when these die. These bone marrow cells ('stem cells') are capable of indefinite division and are able to differentiate into many of the different cell types found in the peripheral blood system. If a mouse is given a dose of radiation of some 2–10 Gy it will cause the inhibition of cell division in the stem cell population of the bone marrow. This will mean that the animals will suffer from an insufficiency of the different types of peripheral blood cells that may eventually be lethal (see Chapter 6). It is possible to save the life of such a critically irradiated mouse by giving it an intravenous injection of unirradiated bone marrow cells. These transplanted cells will repopulate the marrow of the irradiated mouse, divide and provide the vital supply of differentiated peripheral blood cells (see also p. 96). The method devised to assess the radiation survival of bone marrow cells *in vivo* depends on the fact that a fraction of the injected bone marrow cells settle in the spleen of the irradiated mouse. The cells divide there and form colonies or nodules of some 10^6 cells that are clearly visible some 10 days after the injection. Figure 5.3 shows the technical details of the method, which are as follows. A donor mouse is killed and its femur is dissected out. The bone marrow cells are gently flushed out of the central cavity of the bone with a hypodermic syringe containing tissue culture medium. The number of nucleated bone marrow cells in the suspension is counted and varying numbers of them are injected intravenously into a number of recipients. The recipient mice have to be given a dose of radiation to suppress their own bone marrow cells. A fraction of the injected cells will lodge in the spleen. After 10 days these recipients are killed and their spleens examined for colonies.

It takes about 10^5 bone marrow cells to produce about 10 to 15 spleen colonies, so there is approximately one 'colony forming unit' (CFU), as they are called, for every 10^4 bone marrow cells. The vast majority of the nucleated bone marrow cells are seen to be unable to form spleen colonies. There is experimental evidence that each colony forming unit is equivalent to a stem cell and that each colony is derived from the divisions of a single cells.

If the donor mouse is irradiated, a proportion of its colony forming cells will be killed and so if 10^5 irradiated bone marrow cells are injected into recipient mice they will not give rise to 10 to 15 spleen colonies. If it takes 10^5 unirradiated bone marrow cells to give from 10 to 15 colonies per spleen, and if it takes 10^6 bone marrow cells irradiated with approximately 2·5 Gy to give from 10 to 15 colonies per spleen this means that 9 out of 10 of the irradiated colony forming cells have been inactivated.

Figure 5.4 shows the effect of various doses of radiation on the colony forming ability of mouse bone marrow cells. The curve has a small shoulder at low doses and the exponential part has a D_0 of 0·95 Gy.

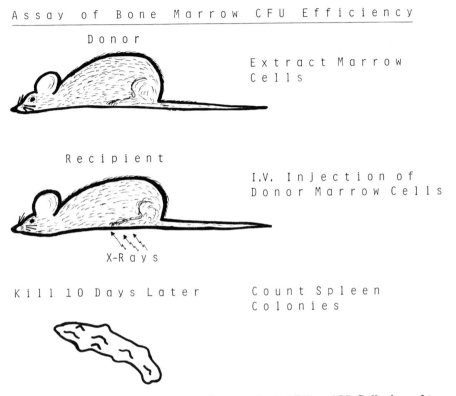

Assay of Bone Marrow CFU Efficiency

Donor — Extract Marrow Cells

Recipient — I.V. Injection of Donor Marrow Cells

X-Rays

Kill 10 Days Later — Count Spleen Colonies

Figure 5.3.   Diagram of the spleen colony method of Till and McCulloch used to assay bone marrow cell survival *in vivo*
Source: J. E. Till and E. A. McCulloch, 1961, *Radiation Research* **13**, 213; courtesy the authors and Academic Press.

The criterion of survival is whether bone marrow cells can divide and produce colonies in the spleens. This criterion is strictly analogous to that used *in vitro* in the Puck plating method since chromosomal analysis has shown that the colonies in the spleen are probably derived from single cells just as the clones on the petri dish are derived from single plated cells.

So, the radiation survival of the bone marrow stem cells *in vivo* is strictly comparable to the mammalian cell survival response *in vitro*. Furthermore, bone marrow cells show the same fluctuating recovery curve as cells *in vitro* (see p. 72). Figure 5.5 shows the effect of fractionated doses of radiation, with the 'survival rate', the ratio of the

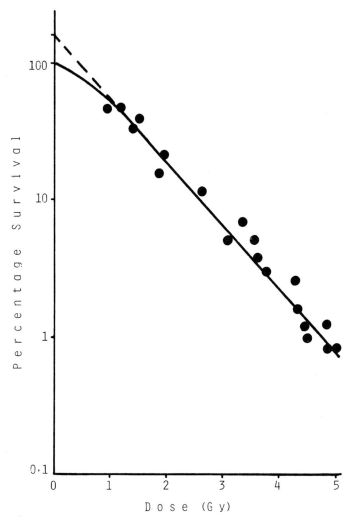

Figure 5.4. Survival curve for the colony-forming ability of mouse bone marrow cells irradiated *in vivo* **with cobalt-60 γ rays**
Source: J. E. Till and E. A. McCulloch, 1961, *Radiation Research* **13**, 213; courtesy the authors and Academic Press.
$D_0 = 0 \cdot 95$ Gy, $N = 1 \cdot 5$.

survival of cells after a split dose to survival after a single dose, plotted against the time between two equal fractions of radiation.

Besides the spleen colony method there are also *in vivo* methods for clonal survival of cartilage, epidermal, spermatogonial and intestinal epithelial cells. Once again a comparison of survival curve parameters shows that there are few systematic differences in radiosensitivity between cells assayed using *in vitro* or *in vivo* methods. There is limited,

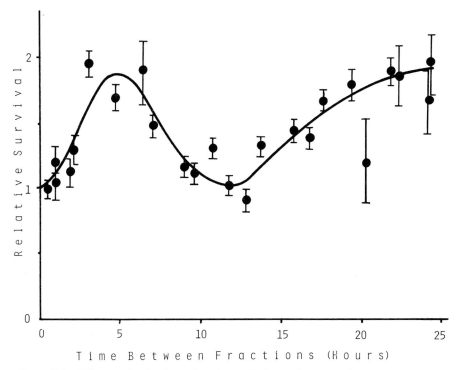

Figure 5.5.   The repair of colony-forming cells from the mouse bone marrow irradiated *in vivo* **with two 2 Gy doses of cobalt-60 γ rays separated by the time intervals shown**
Source: J. E. Till and E. A. McCulloch, 1963; *Radiation Research* **18**, 96; courtesy the authors and Academic Press.
'Survival ratio' is defined as the ratio of the percentage of surviving colony-forming cells after two doses separated by a time interval, *t*, to the percentage survival for the same total dose given as a single exposure.

though suggestive evidence that the survival curves for cells left *in situ* in organized tissues after radiation have larger shoulders (larger D_qs and Ns) than either (*a*) survival curves obtained from single cells in *in vitro* cultures, or (*b*) survival curves obtained from cells *in vivo*, but which require disaggregation of the tissue prior to the *in vivo* assay, for example the TD_{50} method. This indicates an enhanced ability to repair radiation damage in cells which are in close contact with one another in organized tissues.

5.5. Summary

The *in vivo* methods described in this chapter give comparable mammalian cell survival results to those obtained using the Puck *in vitro* plating method. The curves for all the techniques are essentially the same and have a shoulder at low doses of low LET radiation, followed by an exponential part at higher doses. The width of the shoulder region is generally greater in semi-organized tissues *in vitro* (spheroid cultures) and in highly organized tissues *in vivo*. The higher D_q and N values for such survival systems may reflect an enhanced repair capacity that may be related to subtle cell to cell contacts that are disrupted in the single cell assay systems.

Chapter 6
The effect of radiation at the tissue level

6.1. Tissue radiosensitivity

The last three chapters have been concerned with the effect of radiation on the proliferative ability of cells. This is perhaps the most important effect of radiation at the cellular level—a fact that was grasped early in this century. In 1906 Bergonie and Tribondeau were looking into the effect of radiation on the rat testes and they discovered that the dividing (germinal) cells were markedly affected by the radiation, whilst the non-dividing (interstitial) cells appeared undamaged. They formulated a law on the basis of these observations that states that cells are radiosensitive if they are mitotically active, if they normally undergo many divisions and if they are morphologically and functionally undifferentiated. A differentiated cell is taken to be a mature, specialized cell that is unlikely to undergo cell division. So the radiosensitivity of a tissue is directly proportional to its mitotic activity and inversely proportional to the degree of differentiation of its cells.

This law has led to the generalization that actively dividing tissues are 'radiosensitive' and non-dividing tissues are 'radioresistant'. Thus in mammals, the liver, kidneys, muscles, brain, bones, cartilage and connective tissues are often classified as 'radioresistant tissues' since all these tissues in the adult exhibit little or no active cell division and are composed of mature, specialized cells. In contrast, the cells of the bone marrow, the germinal cells of the ovary and testis, the epithelium of the intestine and the skin are all looked upon as 'radiosensitive tissues'. Such generalizations are often misleading since they give the impression that some cell types are more sensitive than others. In general it is not the cell *types* which are either intrinsically radiosensitive or radioresistant, but the cell *processes,* and in particular it is the process of cell division that determines 'radiosensitivity'. There are, however, exceptions to this, such as the oocyte and the small lymphocyte which do not divide but which are nevertheless radiosensitive. The reason for their radiosensitivity is unknown.

When comparisons of radiosensitivity are made between two types of cell the same criteria must be used for both. It is therefore incorrect to say that a liver cell is less sensitive than the epithelial cells of the intestine without stating the criterion of sensitivity used. If one uses the law of Bergonie and Tribondeau one concludes that the liver cells, which rarely divide, are radioresistant, and if after 10 Gy one compares the state of the liver with that of the intestinal epithelium, the former will appear intact, while the latter is damaged—thus confirming the law. What has happened is that the rapidly dividing epithelial cells of the intestinal crypts have suffered damage to their reproductive ability which has caused defects in the structure of the intestinal wall; in contrast, the static, non-dividing liver cells appear morphologically intact. However, if one stimulates the liver cells to divide by, for example, the surgical removal of two-thirds of the liver mass, one is able to use the same criterion of damage in both the liver cell and the epithelial cell. If such criteria as mitotic delay, chromosome damage and the inhibition of DNA synthesis are determined, the doses needed to give the same effect in the liver and the epithelial cells are very similar.

This point, that it is the process of cell division that is radiosensitive, is clearly illustrated when a dose of radiation is given to the whole body of an animal. The cell systems that have a high rate of cell division generally suffer the most damage, while the tissues that show little active cell proliferation suffer proportionately less.

6.2. The modes of death in mammals exposed to whole body radiation

The bone marrow, the intestinal epithelium, the gonads, the lymphocytes and the skin suffer the most damage following whole body doses of radiation. Damage to the bone marrow is known to be the main cause of death in animals following whole body doses of radiation between about 2 and 10 Gy; damage to the intestinal epithelium is known to be the main cause of death of animals following doses between 10 and 100 Gy and radiation damage to the central nervous system is known to be the main cause of death following large doses of radiation, in excess of 100 Gy.

These three modes of death are called the 'radiation syndromes' and they occur predominantly after total body radiation. Although the syndromes are called central nervous system death, gastrointestinal death and bone marrow death, it must be remembered that all organ systems in the body will be damaged following whole body radiation. The three modes of death can be defined not only by the organs and tissues damaged and by the doses required, but also by the time that elapses before death occurs. This is illustrated in table 6.1, which gives the time of death, the organ principally involved and the lethal radiation doses. As the radiation dose increases the survival time decreases; animals die minutes after 1 000 Gy but survive for 3–5 days after 10–100 Gy and for up to 30 days after 2–10 Gy. If the dose to the whole body is below 2 Gy, the bone marrow is able to recover and the animal survives and will not die within 30 days. It will however, die earlier than the unirradiated control animals of an effect called 'radiation lifeshortening' which is discussed in Chapter 10.

Table 6.1. The approximate survival time and mode of death of rodents following whole body doses of radiation

Whole body dose (Gy)	*Approximate time of death after radiation*	*Mode of death*
⟩100	Few minutes to 48 hours	'Central nervous system syndrome'
10–100	3–5 days	'Gastrointestinal syndrome'
2–10	10–30 days	'Bone marrow syndrome'
⟨2	A few weeks before the un-irradiated control animals	'Radiation lifeshortening'

This classification is of course not rigid and all organs are affected by all the doses and the transitions between modes of death are not sharp.

6.3. The central nervous system syndrome

Following whole body doses of 100 Gy and above, most mammals die within 48 hours from what is called the central nervous system (CNS) syndrome. The mean survival time of the animals varies with the dose, but at 1 000 Gy and above the animals die in a matter of minutes, even before the end of the exposure. Irradiation of the head only, provided the dose is large enough, can also produce similar damage in the CNS and a similar acute death.

In brief, the symptoms that such heavily irradiated animals show are irritability, hyperexcitable responses, epileptic-type fits and coma. These symptoms are associated with pathological damage in the nerve cells of the brain and with damage to the blood vessels of the brain. The immediate changes in fluid and electrolyte balance in the brain are due to the rapid radiation-induced permeability changes in the blood vessels. Since the skull does not allow any expansion, loss of fluid from the blood vessels results in a general increase in the fluid pressure inside the brain. These changes in fluid pressure and electrolyte composition may be the cause of the changes observed in the neurons. There is, however, no agreement as to whether the vascular changes cause the neuronal changes or whether the neurons are directly damaged by the radiation. There is probably also a specific radiation effect on the respiratory centre in the brain since animals sometimes die very quickly with no brain damage visible at *post mortem* examination.

In the CNS syndrome the animals are agitated and irritable, but this is quickly followed by apathy, vomiting, salivation, repeated defaecation and diarrhoea. Soon the animals are unable to co-ordinate their voluntary movements (Ataxia) and become disorientated, involuntary rolling of the eyes occurs, tremors and frequent seizures are followed by the final phase of convulsions, prostration, coma respiratory failure and death.

Depending on the dose, the course of the syndrome may last from a few minutes to 48 hours. The CNS syndrome is irreversible and treatment can only be symptomatic, to reduce any distress associated with nervous or gastrointestinal disorders.

6.4. The gastrointestinal syndrome

Following whole body doses between 10 and 100 Gy most mammals die primarily as a result of damage to the cell renewal system of the lining of the intestine. The mean survival time of the animals after radiation is about 3–5 days. Damage to the epithelium in the small intestine with the resultant systemic infection from intestinal flora is the most critical characteristic of the gastrointestinal (GI) syndrome, although damage to the bone marrow cells does play a part. It is possible to get some though not all of the signs of the GI syndrome by irradiation of the whole isolated intestine.

The internal absorptive surface of the mammalian intestine is greatly enlarged by numerous small foldings called villi, which project into the intestinal lumen. At the base of the villi are the rapidly dividing crypt cells that give rise to the cells which move up the villus to become mature differentiated cells. The cells continue to move slowly towards the tips of the villus, and on the way they secrete enzymes and absorb food and water from the intestinal tract. They are eventually 'pushed off' the tip of the villus by the new cells moving up from the crypt.

It is damage to the crypt cell population that is mainly responsible for the GI syndrome. This undifferentiated rapidly dividing population produces the cells that replace the epithelial cells lost from the tips of the villi. Following radiation there is a gradual destruction of the epithelial lining of the crypts. As the radiation-induced degenerative phase progresses the epithelial cells swell, vacuolate, and their nuclei become pyknotic (pyknosis is a degenerative change in a cell characterized by a shrinkage of the chromatin in the nucleus to a dense structureless mass). These cells are prematurely lost from the villi. Since the mitotic process is a radiosensitive process, the mitotic activity of the crypt cells is reduced to zero in about 30 minutes. The duration of this mitotic depression varies with the dose (see figure 6.1) After this immediate, inhibitory phase a transient rise of mitoses occurs at between two and six hours after the radiation, followed by a second period of decreased mitoses. The period of increased mitoses indicates that not all the crypt cells are killed outright by the radiation, some of them are able to divide and form foci of regeneration for a new epithelium. The regenerative waves become weaker and take longer to appear as increasing doses inflict increasing damage (figure 6.1). At doses below 8 Gy in the rat the residual reproductive activity of the crypt cells is sufficient to renew the lining of

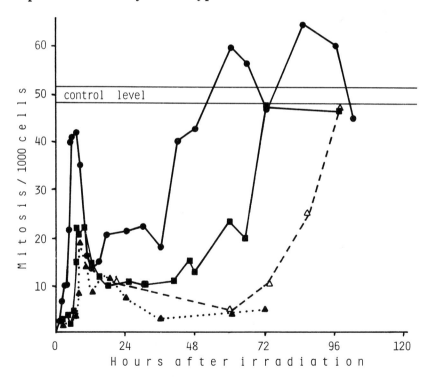

Figure 6.1. Changes in the mitotic count in the epithelial cells of the small intestine of the rat
Source: R. B. Williams *et al.*, 1958, *Journal of the National Cancer Institute* 21, 17; courtesy the authors and the National Cancer Institute.
The band represents the normal range of mitotic activity.
● =3 Gy;   ■ =6 Gy;   ᴀ =8 Gy;   ▲ =9 Gy

the intestine and thus save the animal from intestinal death. At higher doses the rate of epithelial cell loss due to the effects of radiation, combined with the normal sloughing-off processes, is not adequately compensated for by new cell production in the crypts. This failure of the crypt cell renewal system will lead to the destruction of the villi. The very few crypt cells that do survive and attempt to divide do so abnormally. Broken chromosomes, anaphase bridges and fragments (see Chapter 7) indicate that these cells are doomed to die. A few of the abnormal cells produced by the damaged crypt cells may even move into the villi 1–2 days after the irradiation, and these enlarged, vacuolated, misshapen cells are the last remains of the crypt cell population. The crypt cells, no longer capable of division, become detached and their debris collects in the crypt. The loss of the crypt cell divisions means that the villi become progressively shorter and flatter. At the death of the animal, around 3–5 days after the irradiation, the villi may be almost devoid of cells and very nearly flat, although deaths from infection may occur even if the villi are actively regenerating.

The second most important aspect of the gastrointestinal syndrome is the damage to the bone marrow cells.

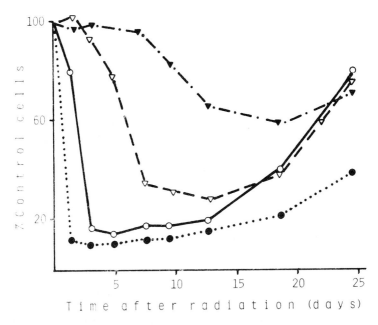

Figure 6.2. The changing pattern of circulating blood cells as a function of the time after moderate (< 10 Gy) whole body radiation
Source: A. P. Casarett, 1968, *Radiation Biology* (C), courtesy Prentice-Hall Inc.
▼ =erythrocytes;   ○=granulocytes;   ▽ =platelets;   ● =lymphocytes.

Circulating blood cells are continually dying and they have to be replaced by new cell production. It is the function of the bone marrow 'stem cells' to maintain the levels of granulocytes, erythrocytes and platelets. The 'stem cells' are very radiosensitive as we shall see in the next section. Following doses of 10 Gy or more there is an almost total arrest of stem cell divisions for the whole of the four-day period of the intestinal syndrome, so that although circulating blood cells will continue to die at their normal rates, no new ones will be produced in the bone marrow; the 'stem cell' renewal system has ceased to function. Since the granulocyte cells are the shortest-lived circulating cells, their numbers in the blood reach very low levels as early as 2–3 days after radiation (see figure 6.2). The rapid fall in the number of circulating lymphocytes in figure 6.2 is a result of direct damage to the mature cells which are exceedingly sensitive and not related to the killing of precursor cells.

The signs and symptoms of the GI syndrome include gastrointestinal pain, a loss of appetite (anorexia), nausea, vomiting, inactivity and inertia and an increasingly severe diarrhoea. The diarrhoea causes marked dehydration which, together with vomiting, causes a loss of weight that in

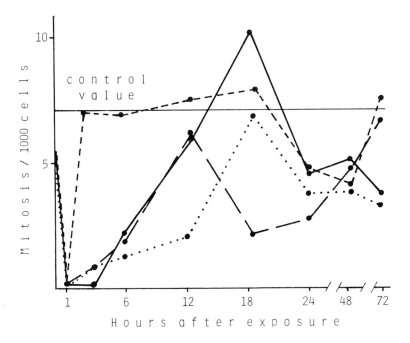

Figure 6.3.   Mitotic index changes in preparations of bone marrow in the rat after whole body doses of cobalt-60 γ rays.
Source: V. P. Bond, T. M. Fliedner and J. A. Archambeau, 1965, in *Mammalian Radiation Lethality,* Academic Press, p. 166, courtesy the authors and publisher.
--●--=1 Gy;   - ●-=3 Gy;   ——●——=5 Gy;   · · ● · ·=7·6 Gy

its turn leads to a state of complete exhaustion and emaciation. The white blood cell count will be very low by three days and the bone marrow will be almost aplastic—devoid of nucleated cells. The dehydration alters the blood volume and the loss of electrolytes causes an alteration in composition of the serum. There may be signs of infection, with bacteria in the blood.

There are three important results of whole body radiation at 10–100 Gy level: (1) fluid and electrolytic losses; (2) the damage to the nutritional system; and (3) infection. Any of these may be sufficiently severe to kill the animal. The failure of the denuded intestine to absorb nutrients and water is reponsible for the severity of the diarrhoea. The dehydration is so severe that the blood becomes increasingly viscous and the balance of salts in the serum is destroyed completely. The loss of the epithelial cell lining will allow bacterial invasion from the intestinal flora and this causes the infection of the underlying tissues and the blood stream. The severity of any infection is enhanced by the fact that there are a reduced number of circulating white blood cells that normally act as a defence against infections.

The death of animals at about 3–5 days after whole body doses of 10–100 Gy is due primarily to the injury sustained by the gastro-intestinal epithelium and by the bone marrow cell renewal system. Damage to these systems causes nutritional impairment, fluid and electrolyte losses and infection.

6.5. The bone marrow syndrome

We saw in the last section that radiation damage to the bone marrow cells formed a significant part of the gastrointestinal syndrome. However, the death of the animals at 3–5 days after radiation is too soon for the full expression of the damage in these bone marrow cells. At doses between 2 and 10 Gy the regenerative ability of the crypt cells in the gastrointestinal tract is great enough to save the animal from GI death. However, animals having received a dose of 2–10 Gy are likely to die within 30 days of the radiation from bone marrow (BM) or 'haemopoietic' death. Haemopoiesis is the process by which the different cell types of the blood are produced. The cells of the bone marrow play a major role in the haemopoietic system. Within a few hours of a whole body dose of 2–10 Gy there is a degeneration or breakdown of the architecture and vascular structure of the bone marrow, and the number of nucleated bone marrow cells is reduced. The reduction in cell numbers is primarily due to the fact that the cells are being inhibited from dividing or killed if they attempt mitosis, so that no new cells are being produced. Besides mitotic inhibition and death many nucleated bone marrow cells are seen to have fragmented nuclei, while others show pyknosis and clumping of the chromatin, which suggests that radiation has killed many cells before they were even able to attempt cell division. This is termed interphase or non-mitotic death. The clearing away of these dead and dying cells is the cause of the rapid fall in the cellularity of the bone marrow. The spaces left by the destruction of nucleated cells tend to fill up with red blood cells that have escaped through the damaged walls of the bone marrow blood vessels (sinusoids). The number of mitoses that are observed in the bone marrow falls to zero within 1 hour of irradiation (figure 6.3). Following this degenerative phase, the cells that have escaped critical radiation injury will attempt to regenerate (figure 6.3). Foci of cells undergoing mitosis are observed (just as was the case in the intestinal epithelium), many of which are abnormal and the cells rapidly die and are eliminated from the marrow.

So, a few gray delivered to the whole body cause profound changes in the nucleated bone marrow population, among which are the vital stem cells that give rise to the peripheral blood cells. A method of assessing the radiosensitivity of the stem cells, the spleen colony method, was given in the last chapter. From figure 5.4 it is seen that a dose of 5 Gy will kill 99 stem cells out of 100. It is not surprising therefore that whole body doses of 6–7 Gy kill an average of 50 per cent of a population of mice within 30 days of the radiation. This dose of 6–7 Gy is therefore known as the lethal dose 50, 30 days ($LD_{50/30}$). Since its exact size varies, not only with the species of animal but also with strain, the values given in table 6.2 are only appoximate.

Table 6.2. The X-ray dose for 50 per cent lethality of various mammalian species in 30 days ($LD_{50/30}$)

Species	Approximate $LD_{50/30}$ (Gy)
Pig	2
Guinea pig	2·5
Man†	3
Monkey	4
Mouse	6·5
Rat	7
Rabbit	8
Gerbil	10

†$LD_{50/60}$ (see footnote p. 102)

Damage to the bone marrow stem cell population is lethal because it is the renewal system of the circulating blood cells. The latter do not divide and are very specialized in their function, soon wear out and have to be replaced. For example, in the mouse the erythrocyte lifespan is approximately 60 days, the granulocyte lifespan is characterized by a half life of some 5-6 hours and the lifespan of the thrombocytes (platelets) is approximately 10 days. The killing of precursor cells is quickly reflected as a fall in the cell counts of the different cell types of the blood (see figure 6.2). Depletion of the peripheral blood elements is responsible for the signs and symptoms that precede death. The signs are anaemia, haemorrhage and infection.

The fall in the number of red blood cells and therefore the haemoglobin content of the blood is the cause of anaemia. Red cells are lost not only as a result of normal ageing and death, but also by escaping through the walls of damaged capillaries.

The fall in the number of platelets (responsible for blood clotting and the prevention of haemorrhage) plays a significant role in the failure to arrest haemorrhage. The haemorrhage adds to the anaemic state and is also more probable where there is radiation damage to vessel walls.

An animal's defence against infection is primarily mediated through its blood and lymphoid systems. Experiments have shown that the radiation-induced fall in the number of circulating granulocytes allows bacterial invasion and proliferation in the blood. This bacterial infection plays a major role in determining the length of time an animal survives following whole body doses between 2 and 10 Gy.

That death, following such doses, is primarily due to bone marrow failure is seen in the experiments in which animals are given a transplant

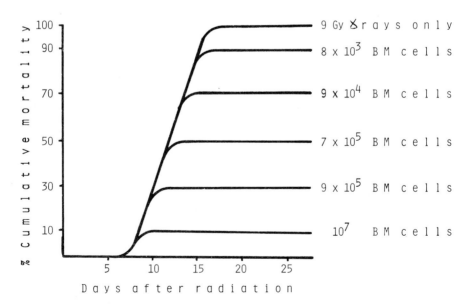

Figure 6.4.   The cumulative mortality of mice given 9 Gy ^{60}Co γ rays and varying numbers of syngeneic bone marrow cells from zero BM cells up to 10^7 BM cells

of unirradiated bone marrow after receiving a lethal dose of radiation. Provided that the number of cells transplanted is large enough, it is possible to rescue the animals from death. Figure 6.4 shows the effect of increasing the number of unirradiated bone marrow cells transplanted into genetically identical, lethally irradiated mice. If 7×10^5 bone marrow cells are transplanted almost 50 per cent of the irradiated mice survive longer than 25 days compared with zero per cent survival if no bone marrow cells are given.

What are the signs of the 'bone marrow syndrome' or 'haemopoietic death'? In the initial period the animals exhibit signs of intestinal disorders such as vomiting and diarrhoea. These soon disappear and from two days onwards the animals are free of symptoms. During this latent phase the two critical effects of radiation occur, the destruction of the bone marrow cells and the resultant depletion of the circulating blood cells. The consequences of red cell loss, platelet loss and granulocyte loss are anaemia, haemorrhage and infection respectively. The latent period is followed by a period of extreme illness. Severe diarrhoea, which is often bloody, heralds serious intestinal disorder that leads to fluid imbalances. The fluid imbalances, together with the haemorrhage that occurs in all the organs and the infection, are the causes of death. Most haemopoietic deaths occur less than 25 days after the whole body irradiation.

6.6. The acute radiation syndromes in man

The three modes of death following acute whole body doses described above for animals have been reported and studied in the following groups of people:

(a) The many thousands of casualties of all ages, victims of the atom bombs dropped over Hiroshima and Nagasaki in Japan in 1945. The radiation dose varied with the distance from the centre of the explosion, and ranged from the immediately lethal level (tens of gray) down to a few hundredths of a gray. At Hiroshima the radiation was predominantly γ rays with a very small neutron component. Besides the uncertainty over the dose received by individual victims, burns, blast and malnutrition complicated the effects of the radiation.

(b) In 1954, 239 Marshall Islanders, 28 United States military personnel and 23 Japanese fishermen were exposed to radiation in the mid-Pacific as a result of fall-out from a thermonuclear device. One Japanese fisherman died.

(c) There have been some 100 accidents involving exposure to high levels of radiation in laboratories, hospitals and industry. These have involved several hundred people of whom 15 died within a few weeks of the exposure.

(*d*) Other people who have experienced more or less severe symptoms of radiation sickness are the more than 1 000 patients treated with whole body radiation (~10 Gy) for leukaemia and other cancers. Some patients have received radiation to suppress their immune responses prior to receiving a kidney transplant.

Whole body acute doses of radiation produce the same spectrum of CNS, GI and BM injury in man as was described for animals. Table 6.3 shows the clinical signs of the three forms of radiation injury in man. Following a whole body dose of about 1·5 Gy X- or γ rays no death is likely, at least in a physically fit adult, but 50 per cent of those exposed will shown signs of gastrointestinal distress with anorexia, nausea,

Table 6.3. Major forms of acute radiation syndrome in man

Time after irradiation	Cerebral and cardio-vascular form (200 Gy)	Gastrointestinal form (20 Gy)	Haemopoietic form (4 Gy)
First day	Nausea Vomiting Diarrhoea Headache Erythema Disorientation Agitation Ataxia Weakness Somnolence Coma Convulsions Shock Death	Nausea Vomiting Diarrhoea	Nausea Vomiting Diarrhoea
Second week		Nausea Vomiting Diarrhoea Fever Emaciation Prostration Death	
Third and fourth weeks			Weakness Fatigue Anorexia Nausea Vomiting Fever Haemorrhage Epilation Recovery (?)

Source A. C. Upton, 1965, *Annual Review of Nuclear Science,* **18**, 495; courtesy the author and publisher.

fatigue and possibly diarrhoea. The cause of this initial or 'prodromal syndrome' is poorly understood.

After a whole body dose of about 4 Gy X- or γ rays, those exposed will suffer 6–7 days of prolonged vomiting and diarrhoea which will abate only to be followed by extreme illness within 3–4 weeks. Haemorrhagic diarrhoea heralds serious intestinal disorder and will cause fluid imbalances which together with life threatening septicaemic infections are the major causes of death. The peak incidence of death corresponds to the 30-day nadir in blood cell numbers, the number of deaths then falls progressively until it reaches zero at about 60 days post-irradiation.

Figure 6.5 shows the probability of BM death in adult humans as a function of X- or γ ray doses. One or two comments are in order. First, it can be seen that over the range 2–5 Gy the probability of death ranges from about 1 per cent to 99 per cent. Second, there is considerable

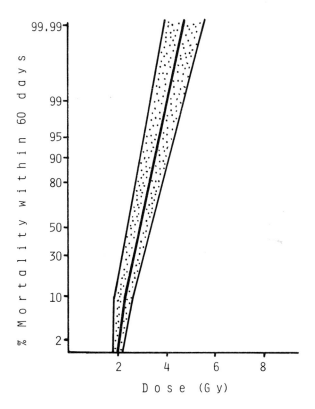

Figure 6.5. **Probability of bone marrow death in adult humans within 6–8 weeks of whole body X- or γ irradiation**
Source: H. Smith and J. W. Stather, 1976, *National Radiological Protection Board Publication NRPB–R52*; modified, courtesy the author and publisher.
The shaded area represents fiducial limits.

uncertainty attached to the slope of the line and to its confidence limits—the $LD_{50/60}$ value† may be as low as 2·5 Gy or as high as 4·5 Gy. Third, this dose–effect relationship is for adults; babies and infants are probably more radiosensitive. Fourth, the curve represents the probability of death in untreated adults. Simple medical treatment such as fluid and blood transfusion, use of antibiotics and in extreme instances HLA-matched‡ bone marrow transplantation might increase the $LD_{50/60}$ from say 3·5 Gy to 5 Gy in adults. The $LD_{50/60}$ for infants and babies is likely to be much less than 3·5 Gy.

Following whole body doses in the intermediate dose range (between 10 and 50 Gy) death in adults occurs within 7–14 days from electrolyte losses, infection and starvation due to an inability to assimilate nutrients. Medically, little can be done to alter the course of the GI syndrome. Similarly, the combined damage to the CNS and blood vessels which causes death in humans within 24–48 hours after whole body doses of more than 50 Gy is medically irreversible.

6.7. The effects of radiation on the gonads and skin of man

Damage to the skin, ovary and testis are an integral and important part of the symptoms of the radiation syndromes in man and they must now be briefly described.

The ovaries in post-natal females contain a population of irreplaceable primary and secondary oocytes at various stages of development. Therefore, radiation, by killing occytes, is able to cause permanent sterility. The data for human ovaries come from reports on radiotherapy patients and from Japanese and Marshall Islanders exposed to fall-out. Single acute doses of 1–2 Gy to both ovaries will cause temporary sterility and the suppression of menstruation for 1–3 years. Acute doses of about 4 Gy will cause permanent sterility.

In contrast to the ovary, the germinal cells of the human testis are at all stages of development. Type A spermatogonia are the primitive stem cells of the testis that undergo cell division and give rise to type B spermatogonia. Type B cells mature into primary spermatocytes which then divide twice, meiotically, to give the spermatids that differentiate into spermatozoa. The spermatogonia are the most radiosensitive cells whereas the more mature elements are radioresistant. So, following moderate doses of radiation, male fertility is not immediately reduced,

---

†$LD_{50/60}$ is the dose in gray required to kill 50 per cent of the population within 60 days of their irradiation.

‡Human lymphocyte antigens, the major group of antigens that provoke strong reactions in tissue transplantation.

since only the spermatogonia are easily killed; the spermatocytes, spermatids and spermatozoa are comparatively unaffected and these surviving cells are able, for a short time, to mature and give rise to spermatozoa. If all the primitive type A spermatogonia have been destroyed, permanent sterility will soon occur. A dose as low as 0·1 Gy produces a low sperm count for up to a year. 2·5 Gy will cause sterility for 2–3 years or longer and 4–6 Gy will cause permanent sterility.

Radiation damage in the skin is a complex of the injuries produced in the tissues of the epidermis, the dermis and subcutaneous layers, and although the principal site of damage is the germinal layer of the epidermis, the most rapid response is shown by the capillary network of the dermis. Dilation of these capillaries and the release of histamine produce a characteristic reddening of the skin (erythema). After moderate doses of 3–8 Gy, erythema may occur within a few minutes but it does not usually last more than 24–48 hours. After this transient erythema there is a latent period before the erythema reoccurs some 2–3 weeks later. This second phase is also due to capillary dilation which is thought to be a response to radiation-induced blockage of small arteries that produces a reduction in the oxygen in the skin. The capillaries dilate in an attempt to compensate for this reduced oxygen supply. The second phase of erythema is accompanied by a loss of the superficial layers of the epidermis which is a result of the radiation killing some of the germinal cells. The skin condition is similar to first degree thermal burns, such as mild sunburn, and may last for a few weeks and then subside. The only scars left are brown patches on the skin.

A dose of 10 Gy causes the second phase of erythema to occur in about a week and it is followed by serious injuries to the skin. The radiation-induced death of the germinal layer causes severe loss of the epidermis (desquamation). This desquamation may be dry flaking or sloughing of the cells ('dry desquamation') or it may be accompanied by blistering, ulceration and fluid exudation ('moist desquamation'). This skin condition is similar to second degree thermal burns, and healing may take weeks and permanent scars are left.

At doses in excess of 50 Gy the injury is severe, the epidermis is destroyed and the dermis and the subcutaneous layers are injured. The reactions appear earlier at these higher doses and the healing of the ulcers and other damage may take years, often healing only to break down again.

The hair-forming cells in the hair follicles are radiosensitive and doses of 3–4 Gy affect hair growth. After such doses the hair soon begins to loosen and it falls out in some 1–3 weeks. Regrowth of hair will occur after these doses, but permanent loss of hair follows doses around 7 Gy.

The lung is the most radiosensitive organ of the thorax. It may be irradiated externally by X- or γ rays, or internally following the inhalation of radioactive particulates. Radiation pneumonitis involves the loss of epithelial cells that line the airways and airsacs, inflammation and occlusion of airways, airsacs and blood vessels, and finally fibrosis. All these effects can cause pulmonary insufficiency and death within a few months of thoracic irradiation. Radiotherapy data suggest that the threshold for acute lung death is probably 25 Gy X- or γ rays, with 100 per cent mortality after 50 Gy to the lung. These doses may be high by a factor of two.

All the organs and tissues of the body can be damaged by radiation but there is no space to describe all these effects. Generally speaking the doses required to produce measurable acute damage in the kidney, liver, pancreas and bone tend to be higher than those mentioned so far in this chapter (see also Chapter 10).

6.8. Treatment of the acute radiation syndromes

There are several types of treatment that increase the chances of survival in animals following acute doses of radiation that have been shown to be of potential use in man.

Nothing can be done to save the life of animals or humans exposed to the high-level doses of radiation that produce irreversible central nervous system death. (In man, treatment is always given to alleviate the pain and agony of the dying victim.)

The symptoms associated with the gastrointestinal syndrome are infection and fluid and electrolyte imbalance. The infection can sometimes be controlled with broad spectrum antibiotics and the electrolytes and fluids can be replaced as necessary. These procedures have sometimes rescued experimental animals from death.

The symptoms associated with the bone marrow syndrome are infection, haemorrhage and a not very severe anaemia. The logical and most effective treatment is bone marrow transplantation, which was discussed on p. 98. Transplantation of bone marrow is more effective between immunologically identical animals, i.e. inbred strains or identical twins. If the donor and the recipient are not immunologically well matched there will be a reaction between the bone marrow graft and the host animal's tissues. This may lead to 'secondary disease', which may be fatal. At present, bone marrow grafts in man are only recommended *in extremis*. The infection, haemorrhage and anaemia of the bone marrow syndrome in man have to be tackled with broad spectrum antibiotics and with massive transfusions of white blood cells, red blood cells and platelets.

Doctors who have had to care for radiation casualties often stress that the cardinal rule is to treat the symptoms *only* as and when they appear.

6.9. *Effects of radiation on the developing organism*

Much information is now available on the effects of radiation on the embryonic and foetal stages of development. Irradiation during development can produce a wide variety of short- and long-term effects. The most important effects include pre-natal and peri-natal death, growth retardation, abnormalities of many of the tissues and organs of the body and the induction of cancer in early post-natal life. The developing organism is especially sensitive to developmental abnormalities during the period called organogenesis—the time of highest proliferative activity and critical cellular differentiation. Much of our information on this topic comes from experimental animal studies, but there are human data and we shall briefly consider both these sources in this section.

It is useful to divide the development *in utero* into three stages since it has been found that the different stages are differentially sensitive to radiation. The stages are:

(1) Pre-implantation, which is the period between conception and the time the early embryo is implanted into the wall of the uterus. In the mouse this is 5 days and in humans, about 6 days.

(2) Organogenesis, which is the period of major organ and tissue differentiation which occurs once the developing embryo is firmly attached; in the mouse this lasts from day 5 post-conception to day 13, and in humans from day 9 to day 60 post-conception.

(3) The foetal period, which is the transition from the embryonic to the post-natal period, and is not clear cut. In the mouse the foetal period lasts from about day 14 post-conception to the day of birth at 19·5 days, and in humans from about day 60 post-conception to the day of birth at about 270 days. By the beginning of the foetal period the major dispositions, shapes and differentiation of the organs and tissues have been achieved and in the foetal period most development is concerned with growth in size.

The effect of radiation at the pre-implantation stage

Pre-natal death of the embryo rather than its malformation is the predominant risk of radiation at the pre-implantation stage. In mice about 50 per cent of pre-implantation embryos die before birth if given doses of about 1 Gy within 24 hours of conception. However, the embryos that survive radiation at this period usually develop normally both before and after birth. The mechanism of pre-natal death probably involves radiation-induced chromosome damage in the cells of the primitive embryo. The embryo degenerates and dies before implantation.

The effect of radiation in the period of organogenesis

Irradiation in this period produces pre-natal and neo-natal deaths. The LD_{50} for pre-natal death is approximately 1–1·5 Gy in rodents irradiated early in organogenesis, although the value of the LD_{50} rises in late organogenesis, and by the foetal period it is similar to that of the post-natal period (about 7 Gy). At this stage the embryo is also sensitive to neo-natal death. For example, neo-natal mortality in dogs is about 80 per cent after 1 Gy given while they are in organogenesis, compared to about 30 per cent in unirradiated dogs. Little neo-natal mortality is observed at doses below 1 Gy. There is little evidence that irradiation *in utero* can cause death in post-natal life.

Besides lethality, radiation given during organogenesis produces a retardation of growth both of the developing foetus and in the post-natal stages and induces a wide spectrum of malformations of many of the structures of the body. The degree of stunting increases with increasing dose and varies with the time during organogenesis that the dose is administered. A dose of 1 Gy or more given after implantation will produce stunting in 100 per cent of offspring that persists into post-natal and adult life.

The peak incidence of gross malformations occurs when the embryo is irradiated early in organogenesis. Abnormalities can be induced in all the major internal and external organs and tissues of the body. It is difficult to summarize the mass of experimental data in this area but one important point emerges and that is the existence of so-called 'critical periods'. A critical period is a time of maximum radiosensitivity for a specific type of abnormality. For example, in one experiment, anophthalmia (absence of eyes) was observed in 25 per cent of mice exposed to 1·5 Gy eight days post-conception, while 1·5 Gy given on day 14 produced no such effects. As the dose increases the critical period for the induction of all such specific abnormalities tends to broaden.

The effect of radiation in the foetal period

The ante-natal radiation sensitivity of all animals decreases during the foetal period for all effects. The LD_{50} dose gradually approaches that of the adult, and growth retardation and the induction of gross abnormalities become rarer, although more subtle microscopic lesions are detectable, especially a widespread tissue hypoplasia, at a dose of 1 Gy or more.

In summary, experimental data show that there is no stage of gestation during which a dose of 0·5 Gy is not associated with a significant probability of inducing damage in mammals, increased mortality during implantation, malformations during organogenesis and cell loss and tissue hypoplasia during the foetal stage.

Moreover, some experiments have shown increased malformations at doses as low as 0·1 Gy, and it is prudent to assume that there is unlikely to be a threshold dose below which no effects will occur.

Human data are scarce and because of insufficient knowledge of conception time, imprecise dosimetry, the smallness of the irradiated groups and the inadequate control populations, precise estimates of the radiation risks for developing humans *in utero* are not possible.

There are no human data for pre-implantation effects.

During organogenesis in humans (days 9–60 post-conception) radiation can induce a variety of malformations of which the best documented are: the impairment of growth, microcephaly (small head size) and mental retardation due to damage to the central nervous sytem (CNS). The peculiar radiosensitivity of the CNS is due to the particularly long period of its development. Proliferation and differentiation of CNS cells begins very early in development and continues into post-natal life. In man CNS neurons only cease to divide at about two years of age.

The lowest doses for which microcephaly has been reported are in the range 0·1–0·2 Gy received at Hiroshima by children irradiated *in utero* before the 18th week of gestation. Microcephaly is often associated with mental retardation and the incidence of both increased sharply with increasing foetal dose in the Japanese victims of the atomic bombs. There are data in the literature about several hundred human embryos that have been exposed to medical therapeutic radiation. A wide variety of abnormalities have been described and although inferences from such information need to be treated with the utmost care they do lead to the conclusion that all mammalian embryos, including those of the human species, exhibit the same general radiation response.

In Chapter 9 we shall consider the evidence that ante-natal diagnostic X-rays in the range 0·002–0·2 Gy can induce childhood cancers (see also below).

However, there remain many areas of ignorance about the effects of radiation on the implanted human embryo. The possibility that low doses (0·05–0·1 Gy) might cause an increase in developmental abnormalities is debatable. Many authorities claim that from a clinical point of view an absorbed dose of up to 0·1 Gy will not cause a significant increase in the incidence of congenital malformations, foetal death or growth retardation, although a significant increase in cancer induction cannot be ruled out. For example, one estimate suggests that between 0 and 1 peri-natal fatality will occur per 1000 births following an embryonic or foetal dose of 0·05 Gy low LET radiation. A similar risk also applies to the induction of severe mental abnormality as diagnosed in adolescence. These radiation risks have to be compared with the natural rates for peri-natal mortality (25 per 1000) and severe metal retardation (4–5 per 1000 adolescents) and with the overall risk in the average pregnancy for the birth of a child with a severe handicap (approximately 30 per 1000 births). In contrast, a dose of 0·05 Gy to the embryo or foetus in the first three months of pregnancy would possibly increase the likelihood of fatal childhood cancer by a factor of ten, to an absolute value of 5 children dying of cancer per 1000 in the first ten years of life. The childhood cancer risk is apparently greater than the teratogenic risk (see also Chapter 9).

Such statements would not receive universal approval among experts, although most would agree that there is probably no threshold dose, i.e. safe dose, for radiation effects *in utero*. All authorities agree that, because of the low probability of their occurrence and the high natural incidence of defects, recognizing radiation-induced damage in offspring irradiated *in utero* gets increasingly difficult with decreasing dose. Nevertheless, since some irradiation of potentially and actually pregnant women is inevitable the various national and international bodies have laid down guidelines for the occupational and clinical exposure of women.

6.10. Summary

The law of Bergonie and Tribondeau states that the radiosensitivity of a tissue is directly proportional to its mitotic activity and inversely proportional to its state of differentiation. It is the process of mitosis that is radiosensitive and it is this fact which confers sensitivity on rapidly dividing cell populations, such as the skin, bone marrow and gonads. In contrast, tissues that contain few dividing cells, such as the brain, muscles, kidney and liver, are termed 'radioresistant'.

Following a whole body dose of radiation the radiosensitive tissues will suffer proportionately more injury than the radioresistant tissues. Small mammals given whole body doses of over 100 Gy die within 1–2 days and at doses greater than 1000 Gy, death occurs within minutes. The cause of death is probably a combination of damage to the central nervous system and the vascular system. Following whole body doses between 10 and 100 Gy, death occurs 4–5 days later from fluid and electrolyte imbalance, infection and nutritional impairment—all due primarily to damage to the gastro-intestinal tract. After whole body doses between 2 and 10 Gy death occurs in less than 30 days and is due to bone marrow failure that causes fatal haemorrhages and facilitates fatal infections.

In man, acute whole body doses of radiation produce the same spectrum of symptoms and the same modes of death as in small mammals, but the doses and times to death are slightly different. The $LD_{50/60}$ for man is about 3 Gy.

All the tissues of the body are affected by whole body doses and the classification of the predominant causes of death above is not rigid, the transition from one type of death to another often being blurred. Besides injury to the central nervous system, gastro-intestinal tract and bone marrow, injuries to the gonads, lungs and the skin are important.

In some instances it is possible to treat the acute radiation syndromes successfully.

Much information is now available on radiation effects on the embryonic and foetal stages of development. The most reliable information on such effects comes from experimental animals, but many of the conclusions are applicable to humans. The main radiation effects are growth retardation, pre-natal and peri-natal death and gross congenital malformations. Growth retardation can be induced by radiation given at any stage of gestation after implantation. The pre-implantation stage is the most radiosensitive for pre-natal death and the stage of major organogenesis the most sensitive for the induction of malformations. The later stages of ante-natal development are the least radiosensitive in all species for all effects. Damage to the central nervous system is the most important effect in man, although the carcinogenetic risks of irradiation at the early stages of pregnancy are probably greater than the teratogenic risks.

Chapter 7
Genetic effects of ionizing radiation

7.1. Introduction

Some 60 years ago American workers using *Drosophila* fruit flies first noted that chromosomes were very susceptible to radiation. These studies were quickly confirmed in many species of plants and animals and doses of much less than 0·1 Gy are now known to cause visible breaks in chromosomes. By the late 1940s the descriptive classification of the different chromosome aberrations was completed. Besides mere description some workers made quantitative studies of the number of aberrations produced by a given dose of radiation. Latterly, attention has been focused on the ability of chromosomes to repair and on the nature of the repair process. Before describing some qualitative and quantitative aspects of chromosome damage it will be useful to present some background information on the structure of chromosomes.

The chromosomes of higher plants and animals are a combination of deoxyribonucleic acid and basic proteins known as 'histones' which are rich in lysine and arginine. The molecular structure of DNA was outlined in Chapter 2. The molecule is thought to be a double helix but this has been seriously challenged recently, particularly by Sasisekharan and his colleagues (see figure 7.1). Their model fits the crystallographic data and has the supreme advantage that one does not have to unwind the molecule to separate the strands. The unwinding problem has always been an almost unsurmountable topological problem for the Watson and Crick model.

The basic DNA molecule is now known to be associated with beads or discs of scaffold proteins (histones) around which the DNA is wrapped some two and a half times. Figure 7.2 shows the packing of these histone beads to form a basic 20–25 nm fibre that can be seen under the electron microscope. This fibre is looped, folded and branched in an irregular

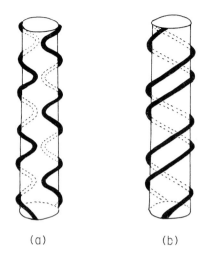

Figure 7.1. **Diagrammatic comparison of (*a*) the Sasisekharan and (*b*) the Watson–Crick models of the DNA molecule**
Source: K. H. Chadwick and L. P. Leenhouts, 1981, *Molecular Theory of Radiation Biology*, Springer-Verlag; courtesy the authors and publisher.

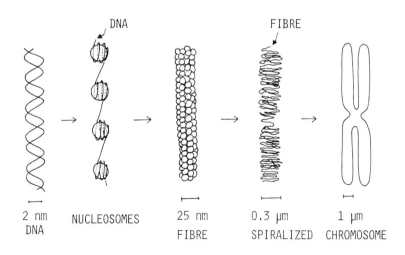

Figure 7.2. **Diagram of the molecular organisation of DNA in chromosomes and its association with histone discs (nucleosomes) that form the single fibre running from one end of the chromosome to the other**
Source: K. H. Chadwick and L. P. Leenhouts, 1981, *Molecular Theory of Radiation Biology*, Springer-Verlag; courtesy the authors and publisher.

fashion to form the metaphase chromosomes that are visible under the light microscope. The differential packing along the choromosome length seems to form the basis of chromosome banding and also accounts for heterochromatic segments. This concept that a single continuous DNA molecule extends from one end of the chromosome to the other is called the unineme concept. The functional integrity of the chromosome relies on the continuity of the single DNA/protein molecule.

The observed effects of radiation on chromosomes can be explained using the theory of chromosome replication given in figure 7.3. This shows a cell with two chromosomes in it, each with a single centromere. As the cell goes through the DNA synthetic phase the chromosomal material will be duplicated, and it is during this phase that each chromosome is divided into two sister chromatids. The chromatids are held together only at the centromere region. At anaphase in cell division the centromere partitions the chromosome and each daughter cell receives one chromatid from each chromosome. The presence of a centromere is essential for the migration of chromosomal pieces to the poles of the cell. 'Acentric' pieces without a centromere fail to move and are usually left in the cytoplasm where they may form micronuclei in one or other of the daughter cells. Pieces with two centromeres, dicentrics, are pulled in opposite directions to form a mechanical 'bridge' between the two sets of chromatids. After cell division the chromatids are known once more as chromosomes. Chromosomes are most readily seen at mitosis, but recent techniques of cell fusion and the phenomenon of

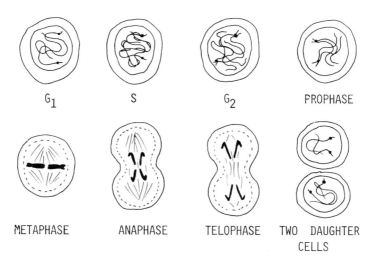

Figure 7.3.   Diagrams of chromosome replication at DNA synthesis (S) phase and the subsequent segregation of chromosomes at mitosis

'precocious chromosome condensation' allow one to visualize single and double chromosomes in the interphase nucleus.

Radiation given to cells before DNA synthesis, i.e. to unduplicated chromosomes, will cause chromosome breaks, while radiation given during G_2 will cause chromatid aberrations and radiation given during synthesis produces a transition of aberration types. The final aberration yield is a combination of radiation breakage and chromosomal repair. The repair may reconstitute the original configuration ('restitution repair'), may involve unison between illegitimate ends to form exchange aberrations ('illegitimate repair') or the break may remain open and give rise to 'terminal deletions'.

We shall first describe some of these basic configurations and then discuss some of the quantitative aspects of the frequency of aberrations versus radiation dose.

7.2. Qualitative aspects of radiation-induced chromosomal aberrations

Figure 7.4 shows the four basic categories of structural change:
1. Interchanges where the interacting lesions occur on different chromosomes.
2. Interchanges when the lesions are within one chromosome. These may be subdivided into:
   (a) inter-arm interchanges when the lesions are on opposite arms of the same chromosome; and
   (b) intra-arm interchanges when the lesions are within one arm of the chromosome.
3. Discontinuities or breaks—the simple severance of the chromosome or chromatid to give an acentric fragment that is *not* associated with any other exchange process.

Interaction of lesions in unduplicated chromosomes or in chromatids may be further classified as asymmetrical or symmetrical exchanges. Asymmetrical exchanges always give rise to one or more acentric fragments that are left behind at anaphase, and not incorporated into either daughter nuclei. Asymmetric exchanges are easily detected, whereas interactions producing symmetrical exchanges do not generally lead to acentric fragments (see figure 7.4) and when there are no great differences in arm length or ratios, symmetrical exchanges are difficult to detect with simple staining methods.

In all types of exchange aberrations the process may be 'complete', i.e. no free or broken ends, or 'incomplete' in which there will be some chromosome parts that are not joined up in new configurations.

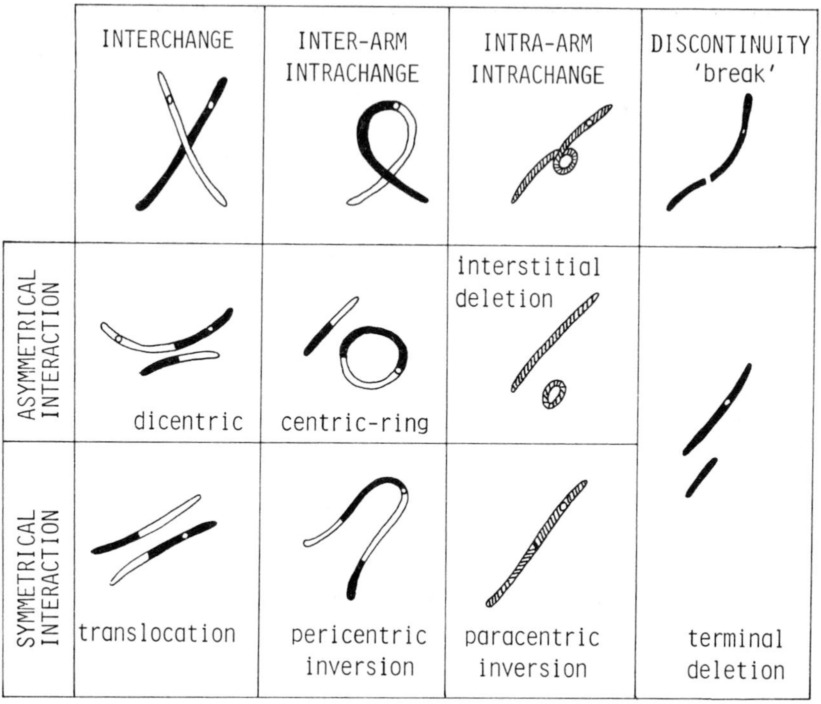

Figure 7.4.   The four basis categories of aberrations
With kind permission of J. R. K. Savage.

Chromatid aberrations show a much greater degree of incompleteness than chromosome-type exchanges (figure 7.5).

It is unnecessary to describe the plethora of aberrations that result from the combination of the four basic types of interactions, plus the fact that they may be asymmetric, symmetric, complete or incomplete, and of either chromosome or chromatid origin.

We shall be concentrating on the fate of the simple chromosome aberrations noted in figure 7.4. Figure 7.6 shows that it is primarily the asymmetrical types that are the problem. These aberrations are termed 'unstable' and are generally lethal to cells; in a rapidly proliferating population they are soon eliminated. They involve one or more acentric fragments that form micronuclei in the cytoplasm of one of the daughters, so that daughter is deficient in genetic information. Asymmetrical interchanges often involve dicentric bridges that usually pose insuperable mechanical separation problems for the cells. Figure 7.6 shows how centric rings may separate at anaphase (cf. a Moebius strip), how they may unravel to form a dicentric loop and how they may interlock. Such dicentrics and rings are lost at a rate of about 50 per cent

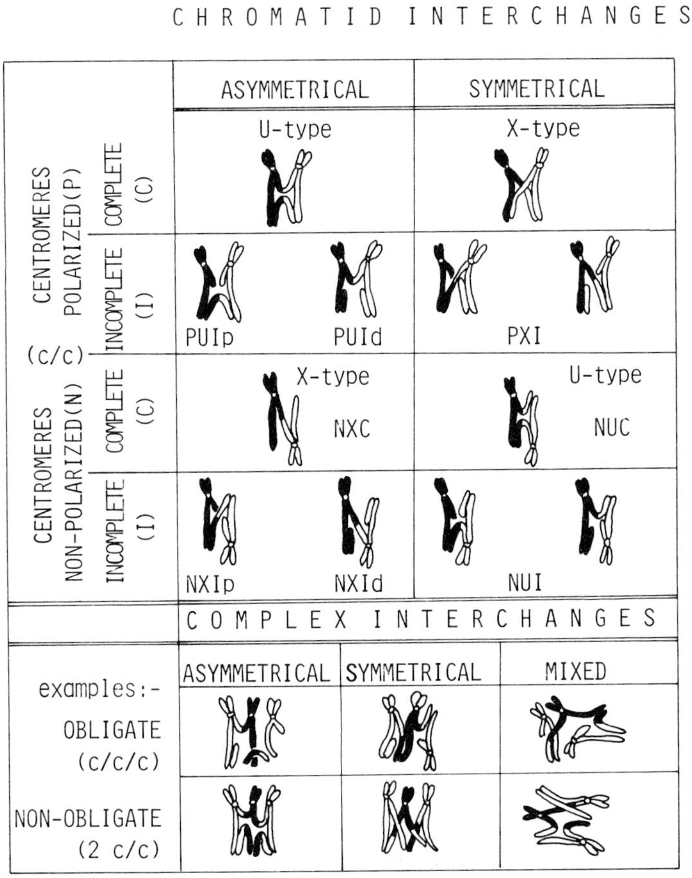

CHROMATID INTERCHANGES

		ASYMMETRICAL		SYMMETRICAL	
CENTROMERES POLARIZED (P)	COMPLETE (C)	U-type		X-type	
	INCOMPLETE (I)	PUIp	PUId	PXI	
(c/c)					
CENTROMERES NON-POLARIZED (N)	COMPLETE (C)	X-type NXC		U-type NUC	
	INCOMPLETE (I)	NXIp	NXId	NUI	

COMPLEX INTERCHANGES

	ASYMMETRICAL	SYMMETRICAL	MIXED
examples:- OBLIGATE (c/c/c)			
NON-OBLIGATE (2 c/c)			

Figure 7.5.   Some examples of chromatid interchanges
With kind permission of J. R. K. Savage.

per division and so reliable scoring of such basic aberrations must be done only in cells at their first post-irradiaton mitosis.

In contrast, symmetrical changes, if complete, do not cause genetic loss or pose any mechanical separation problems at anaphase. Such aberrations are therefore called 'stable' and are potentially more harmful than unstable aberrations, at least in somatic cells, because they can pass through successive cycles. They can only be detected by the recent techniques of chromosome banding.

Chromatid-type structural aberrations arise from irradiation given during or after chromosome duplication at S phase (see figure 7.3). The kinds of lesions mentioned above for chromosomes also occur in chromatids, but the latter's duplex nature means that many kinds of interchromatid interactions may occur. Figure 7.5 gives just a few

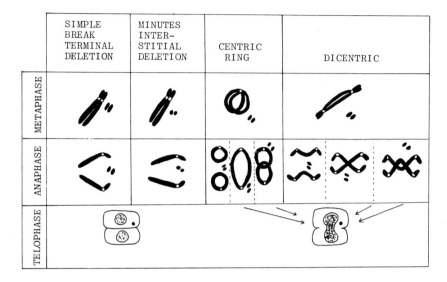

Figure 7.6.   The consequences at mitosis for the four basic classes of asymmetrical chromosome type aberration
With kind permission of J. R. K. Savage.

examples of chromatid interchanges between different chromosomes and also some complex interchanges involving more than two chromosomes. It also illustrates the high frequency of incompleteness to be found in chromatid aberrations.

Finally, it must be mentioned that there is much evidence that a majority of the differentiated cells of both plants and animals normally contain some chromosomal aberrations and these have no apparent effect on either the structure or function of the cell. Most of these 'spontaneous aberrations' are of the chromatid type presumably because they arise as errors or 'miscorrections' during S phase.

7.3.   Quantitative aspects of radiation-induced chromosomal aberrations

Turning from the purely descriptive aspects of chromosome damage we must now look at the relationship between the size of the dose and the yield of aberrations. Experiments with animals and data from irradiated patients have shown that the aberration yield (y) following the exposure of cells to low LET radiation best fits the mathematical function $y = \alpha D + \beta D^2$, where D is the dose, and α and β are constants. This equation is consistent with the hypothesis that some aberrations are the result of the passage of a single ionizing track (single hit event) so that

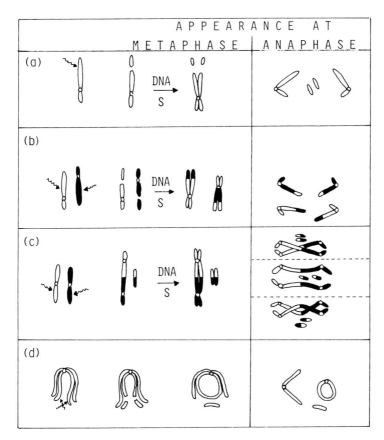

Figure 7.7. Some examples of 1-hit and 2-hit aberrations and their appearance at metaphase and anaphase. (*a*) A simple 1-hit chromosome break; (*b*) A 2-hit symmetrical chromosome interchange; (*c*) A 2-hit asymmetrical chromosome interchange; (*d*) A 2-hit asymmetrical chromatid interchange

the yield is proportional to dose (αD), while other aberrations are produced by two separate tracks when the yield is proportional to the square of the dose (βD^2). Figure 7.7 shows some examples of 1-hit and 2-hit aberrations in both chromosomes and chromatids. The 1-hit aberrations increase in a linear fashion with dose while the more complex aberrations increase in a curvilinear manner (see figure 7.8). The yield of 1-hit aberrations is not expected to be affected by dose rate. If the ionizing events occur at one per second or one per minute, and if each produces a break, then after one minute or one hour at these respective dose rates, the same total dose will have been received and there will be some 60 potential breaks in the chromosomal material. The two lesions required to form the more complex aberrations may come from one or two ionizing tracks, and the actual shape of the dose response curve for 2-hit

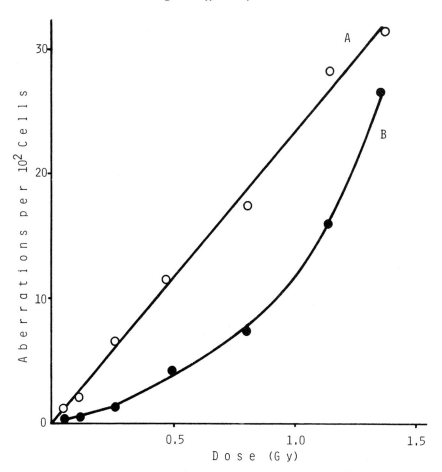

Figure 7.8.   Dose effect curves for X-ray induced chromosomal aberrations
Source: K. Sax, 1945, *Genetics* 25, 41; courtesy the publisher.
Curve A = 1-hit aberrations; curve B = 2-hit aberrations.

aberrations, such as unstable dicentrics, depends on the dose rate and the
LET of the radiation. The dose rate is important because the number of
observed 2-hit aberrations depends on the probability of the first break
and the second break occurring close together in space and time. The two
breaks need to be close enough together in the nucleus for them to
interact and give the 2-hit aberration, and to occur close together in time
so that the first break does not rejoin before the second break is induced.

The LET is important because a single, densely ionizing track of high
LET radiation may easily produce the two lesions needed for a 2-hit
aberration. So, for such high LET radiation as fission spectrum neutrons
the yield of 2-hit aberrations will be linear with dose ($y = \alpha D$).

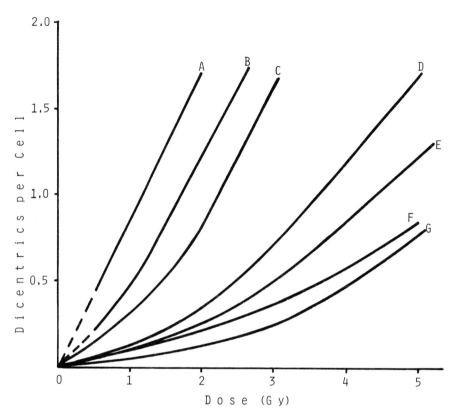

Figure 7.9. *In vitro* **curves for dicentric aberration yields plotted against dose for several qualities and dose rates of radiation**
Source: D.C. Lloyd and R.J. Purrot, 1981, *Radiation Protection Dosimetry,* Volume 1, Nuclear Technology Publishing, p. 19; courtesy the authors and publisher.
Curves A, B and C are for $0 \cdot 7, 7 \cdot 6$ and $14 \cdot 7$ MeV neutrons respectively. Curves D and E are for 250 kVp X-rays at $1 \cdot 0$ Gy min^{-1} and at $0 \cdot 2$ Gy h^{-1}. Curves F and G are for ^{60}Co γ rays at $0 \cdot 5$ Gy min^{-1} and at 0.18 Gy h^{-1} respectively.

These two important points are illustrated in figure 7.9, which gives the yield of unstable dicentric aberrations plotted against dose for several qualities of low and high LET radiations. These aberration yields are obtained from *in vitro* culture of human lymphocytes. This technique is so reliable that it has been used since 1970 in the United Kingdom as a biological dosimeter in all significant radiation accidents. The chromosome damage is assessed in peripheral blood lymphocytes and is a useful check on the dose a person may have received. It can be used to complement the routine film badge or thermoluminescent dosimetry services. Figure 7.9 shows that the yield of 2-hit dicentrics with dose is curvilinear for low LET X- or γ rays but tends to linearity for high LET neutrons. The figure also shows the reduced effectiveness of low dose rate low LET radiation. So, for example, X-rays at 1·0 Gy min⁻¹ are more effective, per unit of dose, than X-rays at 0·2 Gy min⁻¹. Such dose rate effects are indicative of the ability of chromosomes to repair.

The fact that radiation-induced chromosome aberrations can repair has been verified using the split-dose technique. Dividing the dose into two equal fractions and varying the time interval between the first and the second half of the dose allows a proportion of the breaks induced by the first dose to repair before the second half of the dose is given. As the time interval between the first and second dose increases, the yield of 2-hit aberrations will decrease.

Using this split dose technique it has been estimated that the rejoining of the broken ends usually occurs within an hour, although some breaks rejoin in a much shorter time and some take much longer. It should be emphasized that there are a lot of problems associated with fractionation studies on chromosomes. The interpretation of the results, which are often contradictory, is very difficult and many experts are sceptical of ideas based on a simple rejoining picture. There are increasing data to suggest that the complex pattern of chromosome effects seen after protracted and fractionated doses is profoundly influenced by the

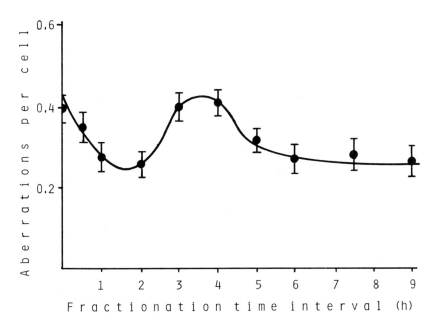

Figure 7.10.   The general relationship between chromosome aberrations (dicentrics and rings) and the time between two doses totalling 2·2 Gy of γ rays
Source: M. S. Sasaki, 1978, in *Mutagen Induced Chromosome Abberations in Human Lymphocytes,* Edinburgh University Press; modified, courtesy the author and publisher.

metabolic state of the cell and this in turn affects the cell's 'DNA-microenvironment'. Figure 7.10 shows the general pattern of fall, temporary rise and secondary fall of the aberration yield as a function of the time between two doses of radiation. The initial fall may be due to the repair of DNA damage and the subsequent rise is thought to be due to the radiosensitization of DNA. The sensitization being in its turn due to an alteration in DNA microenvironment, particularly the repair of the protein component of chromosomes.

At present radiobiologists can only speculate as to the nature and consequence of chromosome aberrations.

There is much evidence that points to DNA strand breaks as a likely primary cause of the damage that leads to exchange aberrations. The structural changes seen at metaphase (figures 7.4, 7.5, 7.6) are of course the final expression of a long chain of biochemical events including abortive attempts to repair or bypass the damage. The repair probably involves not only the DNA repair mechanisms noted in Chapter 2, but also the restitution of the chromosomal proteins. It was noted above that most cells carry a number of 'spontaneous' aberrations that have no apparent deleterious effects. It is only at division that they present problems and even here once the unstable and lethal types have been eliminated by the death of the cells carrying them, the population can tolerate a burden of symmetrical chromosome damage. These remarks refer to somatic cells only; in oocytes, spermatocytes and the cells of the early embryo, chromosome aberrations are of profound hereditary and teratogenic significance.

Recent work has clarified the link between reproductive death of cells at mitosis (see Chapter 3) and chromosome aberrations. An *in vitro* system has been developed that allows one to follow individual cells and to assay them for chromosome damage ('micronuclei') at their first post-irradiation division. Micronuclei in the cytoplasm of a cell are relatively large acentric chromosome fragments that have failed to become incorporated into the genome of the cell. Irradiated cells with and without micronuclei are observed as they grow into colonies. The method has firmly established a 1:1 relationship between the presence of micro-nuclei in a cell and its inability to divide successively and successfully. Cells with micronuclei either do not divide or divide excessively slowly.

7.4. Summary of radiation-induced chromosomal aberrations

Radiation-induced chromosomal aberrations that are observed at cell division are a result of two processes—radiation breakage of either the chromosome- or the chromatid-type, followed by the rejoining of the broken ends in various rearrangements. The aberrations can be divided into 1-hit and 2-hit types. The frequency of simple 1-hit breaks is approximately linearly related to dose and independent of dose rate, while the more complex 2-hit breaks increase in frequency more rapidly than the single power of the dose. The yield of 2-hit aberrations is related to the rate at which the radiation is delivered. Furthermore, 2-hit aberrations show a split-dose effect, the rejoining of the breaks induced by the first dose occurring before the second dose is received, which causes a reduced yield of 2-hit aberrations relative to the yield following a single dose of the same total size. The interpretation of fractionation studies is complex and involves DNA repair, cell metabolism and the DNA microenvironment. There is increasing evidence to link DNA strand breaks, chromosome aberrations and the reproductive death of mammalian cells.

7.5. Genetic mutations

Genetics is the study of inheritance and in particular of genes, which are the factors of inheritance. They are responsible for the infinite variety of structure and function to be found in the animate world. The genetic material is believed to be the deoxyribonucleic acid molecule. DNA complexes with proteins to form chromosomes. The information carried by the genes, which are linearly arranged along the chromosomes, is in the form of the triplet base code of the DNA molecule. At cell division there is a separation of the chromosomes so that each daughter cell receives the same genetic information as is present in the parent cell (see figure 7.3).

Most cells contain two sets of chromosomes that contain matched pairs of genes. A dominant gene is one that can show its effect when it is carried on a single chromosome. A recessive gene can only show its effect if both chromosomes carry it. A dominant gene will override the effect of its recessive partner. For example, the gene for greenness (G) in peas is dominant over the gene for yellowness (g). In order to have green peas, the cells of the plant must either have GG or Gg as a gene complement; whereas to have yellow-coloured peas, the gene complement must be gg. Some genes show incomplete dominance in which there is a blending or co-operative effect between pairs of genes, rather than the suppression of the recessive gene by the dominant gene. Thus if the gene for greenness were not dominant over the gene for yellowness then the combination Gg might yield yellowish-green peas.

All the cells of a given plant or animal are believed to have the same genetic information. This statement is generally true but very, very occasionally changes (mutations) occur in the genetic material. A mutation in a cell means that it no longer has the same genetic complement as neighbouring cells. Mutations may increase, decrease, or qualitatively alter the expression of a gene. They are normally divided into two categories: gene or point mutations which are invisible and involve single gene changes; and chromosome mutations (or aberrations) which are visible with cytogenetic techniques and which involve changes in large sections of the chromosome or even the addition or loss of whole chromosomes.

The importance of mutations is not so great in the somatic cells of the body as in the germ cells. A mutation in a somatic cell may result in the malfunctioning or even the death of that cell or any of its descendants, but since there are so many millions of cells in any organ, the effect of one or two mutants will be insignificant. However, somatic mutations have been suggested as the basis of cancer and of the phenomenon of ageing, and these topics are discussed in Chapter 9 and Chapter 10 respectively. In contrast, mutations occurring in the germ cells may have disastrous effects on the offspring. Mutations occurring at any stage in the development of the ovum and sperm or in the fertilized ovum (zygote) are very likely to lead to the death of the progeny or at least produce seriously defective offspring. The discussion to follow will be concerned only with the effects of mutations in the germ cells and their effect on the progeny.

Gene mutations may be classified as dominant, recessive or sex-linked. A dominant mutation is one that appears in the immediate offspring, as a result of a change in the germ cells of either parent. A recessive mutation will only appear in offspring that have received the same mutation from both parents. A recessive mutation occurring in only one parent's germ cells cannot show itself in the immediate offspring since two matched recessive genes are required before their effect becomes apparent. Eventually, however, in future generations individuals may arise having received the same recessive gene from both parents. Both recessive and dominant mutations may be lethal or they may cause a whole spectrum of visible and invisible effects.

Mutations of the third type are those that appear in the sex chromosomes. Sex is determined by a single pair of sex chromosomes. In mammals the female has two similar sex chromosomes called XX, while the male has dissimilar sex chromosomes, one X and one Y. During the development of the germ cells, all the ova of the female receive a single X chromosome, while half the male's sperm receive an X and the other half a Y chromosome. If the fertilization of an ovum by a sperm gives rise to an XY zygote it will develop into a male, if the result of fertilization is XX, the zygote will develop into a female. The Y chromosome carries very little genetic information, and consequently almost all the genes on the single X chromosome of the male will be expressed, i.e., will appear as dominant genes. Therefore any mutation that occurs in any X chromosome is likely· to be dominant in the XY male. A recessive mutation occurring in the X chromosome will have to be present in both sex chromosomes to show its effect in the female (XX). Like dominant and recessive mutations sex-linked mutations may be lethal or they may cause any number of changes in structure and function of tissues and organs.

It is well known that the mutations that have occurred in the course of evolution have produced the individuality of the different species of plants and animals. Evolution is the process of gradual change in which new species arise better adapted to their environment than their predecessors. Through the process of spontaneous mutations a species is able to adapt to a changing environment. The failure of a species to do so will eventually lead to its extinction and perhaps to its replacement by a more perfectly adapted one. Spontaneous or naturally occurring mutations are the prime movers of the evolutionary process. The causes of such mutations are not fully understood, although it is known that chemicals, including the sex hormones, ultraviolet and ionizing radiation, and high temperature can all increase the mutation rate in plants and animals. The relative contribution of these factors to the spontaneous rate of mutations is not known.

It is generally believed that most mutations are harmful for the simple reason that over the millions of years of evolution, the mechanisms of cells and organisms have become so intricate and attuned to their function that the slightest change is likely to be disastrous. Also, most of the advantageous mutations will already have been selected and become an integral part of the cell. It is much easier to damage a sophisticated mechanism than it is to improve its efficiency.

The vast majority of the world's radiation geneticists would subscribe to the statement that "Any increase in the amount of ionizing radiation to which human populations are exposed is expected to bring about a proportional increase in the frequency of mutations". What is the evidence for such a statement? In the first half of this chapter we saw that radiation causes all sorts of visible chromosomal defects and in this second half we will discuss the evidence in man and animals that radiation causes heritable genetic damage.

7.6. Chromosome mutations following irradiation

The genetic effects of radiation are the result of gene mutations and chromosome aberrations. Structural and numerical errors in human chromosomes are believed to constitute the major part of radiation-induced genetic damage; even a significant proportion of specific locus or point mutations are now thought to be the result of minute chromosome deletions.

Since there are virtually no useful human radiation genetics data, the only way to evaluate the genetic risk in man is to make a number of assumptions and apply the results obtained from experimental animals. The estimated mutation rates per unit of radiation can then be compared with the natural frequency of genetic disease in man. We shall begin with the latter—the estimates of the rates of spontaneous incidence of various kinds of genetic disease in man. Table 7.1 gives the best estimates made in 1977 by the UN Scientific Committee on the Effects of Atomic Radiation (UNSCEAR).

Table 7.1. Natural incidence of genetic disease in man per 10^6 live births

Gene Mutations	
Dominant/X-linked	10 000
Recessive	2 500
Irregularly inherited:	
Malformation	24 000
Constitutional and degenerative	64 000
Childhood neoplasms	1 400
Chromosome Alterations	
Structural—balanced	1 900
—unbalanced	500
Numerical—sex	2 000
—autosomal	1 200
Total	**107 500** (~10% of all live births)

From table 7.1 it is seen that about 10 per cent of people will have more or less serious trouble some time in their lives due to diseases or defects wholly or partly genetic in origin, nine out of ten of these defects being 'irregularly inherited conditions'. It needs to be stressed that we do not know what percentage of these 'spontaneous' genetic mutations are a result of some man-made hazard such as radiation.

Each of the categories in table 7.1 contains a wide variety of defects and diseases that affect most of the systems and tissues of the body. The type of defects produced by autosomal diseases include anomalies of nerves, eyes, skin, cartilage, skull and facial bones. Of the X-linked diseases, haemophilia, progressive muscular dystrophy, hypogammaglobinaemia, agammaglobinaemia and colour blindness are the most common. Of the recessive diseases, phenylketonuria, cystic fibrosis, albinism, deafness and metabolic disorders are prominent. Of the chromosomal disorders Down's syndrome (mongolism, trisomy for chromosome 21) is predominant with sex chromosome anomalies, Klinefelter's (XXY) and Turner's (XO) syndromes, much less frequent. By far the largest groups of defects are the congenital malformations, and there are scores of these of which the best known are probably spina bifida, hydrocephalus, cataracts, strabismus, septal defects in the heart, cleft palate and lip, club-feet and -hands and congenital dislocation of the hip. The multifactorial disorders include diabetes, mental retardation, epilepsy, schizophrenia and myopia.

The only genetic marker that one can easily check in individual men or women is the chromosome constitution or karyotype—any numerical or structural abnormalities can be easily identified with light microscopy. The normal or 'euploid' number of human chromosomes is 46 (22 pairs of autosomes and 1 pair of sex chromosomes); any number other than 46 is called aneuploid. To date about 60 000 infants throughout the world have been karyotyped and 336 of them were found to be abnormal (0·6 per cent). Of the 0·6 per cent, 0·22 per cent showed sex chromosome anomalies (XYY, XXY, XO), 0·14 per cent autosomal trisomy (three instead of the correct pair of chomosomes), 0·19 per cent euploid structural rearrangements and 0·05 per cent aneuploid structural rearrangements.

From this information the natural chromosome mutation rates can be derived for the different types of abnormalities. For example, the spontaneous mutation rate for errors in the number of autosomal chromosomes is $6·7 \times 10^{-4}$ per gamete per generation. This means that about seven gametes (oocytes and spermatozoa) in ten thousand will have the wrong number of autosomes. The rate for the other categories of chromosome mutations are of the same order of magnitude—between 2 and 7.5 per 10^4 per gamete per generation.

The harmfulness of chromosome aberrations is clearly seen in relation to spontaneous abortions, where they occur very frequently. In a number of surveys about 66 per cent of aborted embryos, arrested in their development less than 8 weeks after conception, showed some kind of chromosome abnormality. This compares with an incidence of 0·6 per cent chromosome abnormalities seen in new-born babies (see above). The commonest kind of abnormality in such abortions, accounting for 50 per cent of all aberrations, is trisomy for one of the autosomes; other abnormalities are monosomy, triploidy and tetraploidy.

Radiation is known to cause all types of chromosome mutations in experimental animals but there is still little evidence of a link between parental irradiation and an increase in abortions, still-births and so on in man. Some surveys have shown an association between a mother's irradiation history and the likelihood of her producing a chromosomally abnormal foetus. However, studies have failed to show unequivocally any link between maternal irradiation and the risk of producing a mongol child. Similarly no firm link has been made between parental exposure and mortality rates among children born to A-bomb survivors. These negative findings are, however, useful in that they allow calculations of values for the lower limits of a 'doubling dose' (see p. 129).

7.7. Chromosome translocations following irradiation

The translocation of pieces of chromosome from one chromosome to another can be induced at all stages of spermatogenesis and has been extensively studied in male mice. For technical reasons data for female germ cells remain relatively meagre. The different stages of oogenesis and spermatogenesis often show significant differences in sensitivity and dose response relationships. For example, after spermatogonial X-irradiation in mice the frequency of translocations increases with doses up to 7 Gy and then declines at doses greater than 7 Gy, whereas no such decline is seen with spermatozoal X-irradiation. Both fractionated and low dose rate X- and γ rays are less efficient at inducing translocations than single acute doses. At dose rates as low as $0 \cdot 03$ mGy min^{-1}, ten times fewer translocations are induced in spermatogonia than at 1 Gy min^{-1}. Fractionation studies of irradiated human spermatogonia have shown them to be three times more sensitive to translocation induction than mouse spermatogonia. However, since the data for other primates such as rhesus monkeys and marmosets are so variable it is difficult to make risk evaluations from the small amount of human data.

The major effect of chromosome translocations in experimental animals is the induction of lines of sterile and semi-sterile males.

7.8. Dominant lethal mutations following irradiation

Dominant lethal mutations are point or chromosome mutations that occur in parent germ cells and cause embryonic death. Male or female animals are irradiated and then mated with unirradiated partners and the induced dominant lethals measured by pre-natal methods. These involve examining the uterus of the mated females and counting corpora lutea, living and dead embryos or assessing the ratio of dead implants to total implants. Irradiation causes an increase in post-implantation mortality in all laboratory animals, although there are striking differences in sensitivity not only between different species (hamsters, guinea pigs and mice) but also between different strains of mice. The relationship between the dose and the induction of dominant lethals varies in different studies, some showing a linear response while other work shows exponential dose response curves.

7.9. Gene or point mutations following irradiation

Point mutations involve a change in a single gene and may be dominant, recessive or sex-linked. Most experimental radiation genetics has been done with *Drosophila* and mice, and mutations affecting all parts of the body have been induced. The mutations may have almost imperceptible effects or they may be lethal. In early *Drosophila* work, radiation-induced mutation frequency was shown to be linearly related to dose down to 50 mGy. Extrapolation of the line in figure 7.11 to the axis gives the frequency of spontaneous mutations. The figure indicates that genetic point mutations show no 'threshold dose', i.e. all doses, however small, are effective in causing mutations. The linearity of the effect with dose is analogous to that for single chromatid and chromosome breaks. The linearity suggests that the rate at which the dose is given will not affect the mutation yield and that the possibility of repair of mutations does not exist.

When these early conclusions from the *Drosophila* work were extrapolated to man they suggested that however small the dose and however low the dose rate might be, a cumulative genetic hazard existed, since biological repair seemed impossible.

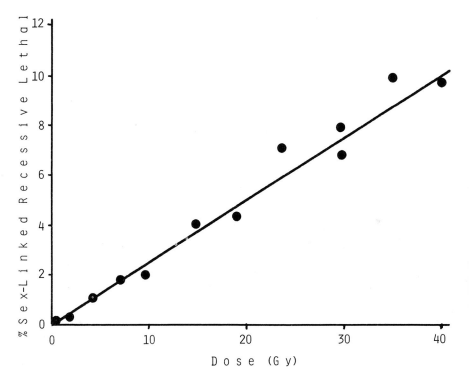

Figure 7.11. The relationship between radiation dose and the frequency of sex linked recessive lethals induced in *Drosophila* spermatozoa
Source: C. E. Purdom, 1963, in *Genetic Effects of Radiation*, George Newnes Ltd., courtesy the author.

Recent work with experimental mice has radically altered this picture. These studies suggest that repair processes do intervene between the induction of primary damage and its final expression and that dose protraction or fractionation significantly lowers the number of mutations induced. It is now known that exposure rates less than 8 mGy min^{-1} (and down to 0·007 mGy min^{-1}) to mouse spermatogonia produce one-third of the number of mutations produced by more acute dose rates. Similarly, certain dose fractionation regimes produce fewer mutations than single large doses.

It is beyond the scope of this book to detail the results of the large scale breeding experiments (sometimes involving more than 100 000 animals) that have led to the massive body of evidence on the mutagenic effects of radiation in the mouse. Many people now feel that these data provides a firm basis for the scientific estimation of the genetic risks of radiation in man. Needless to say, there are those who stress the differences between mice and men and point to the diversity of results found in other mammalian species. In the next section we shall ignore the sceptics and deal with the attempts that have been made to evaluate human radiation genetic risks.

7.10. Evaluation of radiation genetic hazards in man

There are virtually no relevant human radiation genetic data and so risk evaluations are based on information from experiments, and the majority of this comes from work with the laboratory mouse. Since most genetic doses likely to be received by human populations are at very low dose rates (environmental radiation is 1 mGy yr^{-1}; $1·9 \times 10^{-6}$ mGy min^{-1}); the most relevant mouse work is that involving oocytes and spermatogonia irradiated at appropriately low dose rates.

There has been a recent suggestion that there is a relatively simple linear relationship between the DNA content of an organism and the radiation-induced gene mutation frequency. This would imply that the extrapolation of such rates from micro-organisms to man were valid and would be most useful in assessing risks. However, an international committee of genetic experts convened by the United Nations (UNSCEAR) did not accept this hypothesis and maintained that any such extrapolations are unwarranted. Nevertheless, the Committee felt the need to make some evaluation of the genetic hazards to man and to do this they have used two methods—a 'direct method' and a 'doubling dose method'.

The UNSCEAR 1977 genetic risk estimates for man for the direct method are given in table 7.2 and for the doubling dose method in table 7.3.

Table 7.2. Risk of induction of various kinds of genetic damage in man per 10 mGy at low doses or low dose rates of low LET radiation

Endpoint	Expected rate of induction per 10^6 gametes resulting from irradiation of		Expression in first generation per 10^6 births
	Spermatogonia	Oocytes	
1. Autosomal mutations	60	—	20
2. Dominant visibles	v. low	—	—
3. Skeletal mutations	4	—	—
4. Balanced reciprocal translocations	17–87	low	low
5. Unbalanced products of 4	34–174	—	2–10
6. X chromosome loss	v. low	low	low
7. Other chromosome anomalies	—	—	—

Data from *Report to the United Nations Scientific Committee on the Effects of Atomic Radiation (UNSCEAR) 1977,* UN Report A/32/40.

Table 7.3. Estimated effect of 10 mGy per generation of low dose, low dose rate, low LET radiation on a population of a million live-born individuals

Disease category	Current natural incidence	Effect of 10 mGy per generation	
		First generation	Equilibrium
Autosomal dominant and X-linked disease	10 000	20	100
Recessive disease	1 100	slight	v. slow increase
Chromosomal disease	4 000	38	40
Congenital abnormalities, anomalies expressed in later life, and constitutional and degenerative disease	90 000	5	45
Total	105 000	63	185
% of current incidence		0·06	0·17

Data from *Report to the United Nations Scientific Committee on the Effects of Atomic Radiation (UNSCEAR) 1977,* UN Report A/32/40.

The direct method expresses risks in terms of the amount of genetic damage expected per gamete per unit dose of radiation. This involves using mouse radiation-induced mutation rate data and multiplying them by appropriate figures for the human genome to arrive at a likely human mutation rate. But since there are no easy ways to estimate precisely the number of gene loci in the human genome the calculations made by this multiplication method remain questionable. For example, using the direct method the 1977 UNSCEAR Committee made the following estimates of the risks to man of the induction of dominant mutations.

Chronic irradiation of mouse spermatogonia induces dominant visible mutations in experimental mice at a rate of $1{\cdot}0\times10^{-5}$ per Gy per gamete. Dividing this rate by the number of gene loci in the mouse at which dominant visible mutants have occurred, about 75, we obtain a per locus rate of $1{\cdot}3\times10^{-7}$. Multiplying this figure by the assumed number of gene loci which determine dominant traits in man, about 1000, we obtain an overall rate for dominant mutations of $1{\cdot}3\times10^{-4}$ per Gy per generation.

However, the 1977 UNSCEAR committee felt unable to use this figure because of the limitations and uncertainties of this approach and, as can be seen from table 7.2, there are numerous gaps where quantitative estimates are not given. Indeed, even the values given are fraught with uncertainty.

The doubling dose method is a way of expressing risks in relation to the natural incidence of human genetic disease. It is defined as that dose which will exactly match the natural rate of genetic mutations. The method assumes firstly that one knows the natural incidences of the different classes of genetic disease in man, secondly that one knows to what extent these incidences are maintained in the human population by mutations and thirdly and most importantly, it assumes that there is some proportionality between the spontaneous mutation rate and induced rates so that the incidences will increase in the same way after radiation.

The doubling dose, estimated by the genetic committee of UNSCEAR to be 1 Gy, is derived from various mouse experiments and from the mortality data for children born to A-bomb survivors. This doubling dose has been used, together with the data in figure 7.1 for the natural incidence of different classes of human genetic disease, to calculate the expected, additional, genetic disease caused by 10 mGy of low dose, low dose rate, low LET radiation in a population of a million live-born. This is given in table 7.3.

Some explanation is needed of the figures in table 7.3. The first column of figures, with slight modification, is from table 7.1 for the natural incidence of different types of genetic disease. It is assumed that all autosomal dominant and X-linked diseases will increase proportionally with dose, so a doubling dose of 1 Gy will by definition induce 10 000 cases of such disease per million live-born and 10 mGy will induce 1 per cent of this figure—100 cases. However, only 20 per cent of these are likely to arise in the first generation.

Of the $0{\cdot}6$ per cent chromosome abnormalities seen in new-born babies (see p. 126) two-thirds ($0{\cdot}4$ per cent) are considered as the incidence causing chromosome disease and wholly maintained by mutations. Therefore, if the doubling dose is 1 Gy and induces 4000 cases of such disease per million live-born, 10 mGy will induce 40 cases, and most of these will appear in the first generation.

As we have seen, the majority of human genetic diseases are irregularly inherited and caused by complex multifactorial and congenital factors, but no one knows what proportion of these will respond to radiation. The UNSCEAR 1977 committee have assumed that 5 per cent of such disorders are inducible by radiation, so the doubling dose of 1 Gy would induce 4500 cases (5 per cent of a natural incidence of 90 000) and 10 mGy would therefore induce 45 cases of which only 5 are expected in the first generation.

Overall, UNSCEAR 1977 estimates that 10 mGy would induce, in a population of one million live-born, a total of 63 cases of genetic disease in the first generation—0·06 per cent of the natural incidence (10·5 per cent)—and a total of 185 cases altogether—0·17 per cent of the current incidence. It must be stressed that these estimates are very uncertain and depend on somewhat fragile assumptions concerning "(A) the comparability of natural and radiation-induced mutations and (B) the rate at which newly arisen mutations are eliminated from the population".

The kind of genetic data referred to in this chapter are used by the international regulatory bodies in their evaluation of the overall risk of radiation (see Chapter 12). The 1977 report of the International Commission on Radiological Protection (ICRP 26) notes that the basis for inferring risks of hereditary damage in man still rests on observations in mice although these have recently become increasingly firm. The commission concludes that the genetic harm from radiation is likely to be less than the 'detriment' due to somatic injury—mainly cancer induction (see Chapters 9 and 12). The risk of serious hereditary ill health within the first two generations following uniform whole-body low LET irradiation of either parent is about 10^{-2} Sv^{-1} (sievert) and the additional damage to later generations is of the same magnitude. In radiation protection the average risk factor for hereditary effects for exposed individuals will be substantially lower when account is taken of the proportion of the exposures that are of no genetic significance. This average risk for hereditary effects in the first two generations is estimated to be 4×10^{-3} Gy^{-1} and the additional damage to later generations is of the same magnitude.

Nevertheless, it is clear that any increase in radiation levels is undesirable because of its implications for future generations, but as the UNSCEAR report stresses, "the proper use of radiation in medicine and industry is important for the health of the individual and the welfare of the community".

7.11. Summary

Radiation has been shown to cause a whole spectrum of heritable mutations in experimental animals. Dominant, recessive and sex-linked lethal mutations and visible chromosome defects are all known to follow irradiation. These lethal mutations cause the death of developing embryos, premature abortions of foetuses and stillbirths.

Besides lethal mutations, a whole gamut of radiation-induced gene and chromosomal mutations have been described in animals that can cause a wide range of more or less severe malformations of the external and internal organs and tissues of the adult animal.

Early work, in particular with *Drosophila*, suggested that the yield of gene mutations is linearly related to the dose and that the rate at which the radiation is given has no effect on the yield of mutations. It also suggested that no threshold dose exists for gene mutations and that however small the dose may be it has a mutagenic effect.

Recent work with experimental mice has radically altered this picture and fractionation and dose rate studies indicate that repair of pre-mutational damage is possible. For example, exposure rates of less than 8 mGy min^{-1} to mouse spermatogonia produce one-third of the number of mutations induced by more acute dose rates.

There are so few human radiation genetics data that risk evaluations have to be based primarily on mouse data and of the two methods used, the direct method and the doubling dose method, the latter is the more instructive. It is a way of expressing risks in relation to the natural incidence of human genetic disease. Assuming a value of 1 Gy for the doubling dose, UNSCEAR 1977 estimates that 10 mGy would induce 185 cases of genetic disease in a population of one million live-born. This figure is 0·17 per cent of the natural incidence of genetic disease in man. Such an estimate is subject to many uncertainties. The ICRP Report 26 (1977) considers that the genetic harm from radiation is likely to be less than the detriment due to somatic injury. Despite this assurance any increase in environmental levels of radiation is highly undesirable.

Chapter 8
Some factors which modify the biological effect of radiation

8.1. Introduction

The complexity and number of events between the initial absorption of radiation energy and the final expression of biological damage are so great that numerous modifications are possible. Physical, chemical and biological factors can modify the amount of radiation-induced damage.

8.2. Physical factors that influence the effect of radiation

Throughout this book we have seen that as the size of the dose increases there is generally an increase in biological damage—the larger the dose the larger the effect. This statement is only partly true because it is necessary to consider dose rate, dose fractionation and the quality (LET) of the radiation. In general, the effect produced by a given dose of radiation decreases as the dose rate falls; for example in Chapter 7 we saw that as the dose rate decreased, the yield of 2-hit chromosome aberrations per unit dose decreased (figure 7.10).

Figure 8.1 shows the effect of dose rate on cell survival following low LET X-rays and high LET neutrons. The effect is interpreted in terms of repair—the lower dose rates allowing greater survival because there is more time available for any repair to occur. Dose rate effects are more noticeable for low LET than high LET radiations because the larger shoulder on the survival curve for low LET radiation indicates a greater potential for the repair of sub-lethal damage. Similarly, the splitting of a dose into small fractions allows repair between the fractions and so less biological damage is observed in such regimes than for a single large dose of the same total size (see Chapter 4). Indeed, the low dose rate effect is equivalent to exposure to very many small fractions (see dotted Curve A,

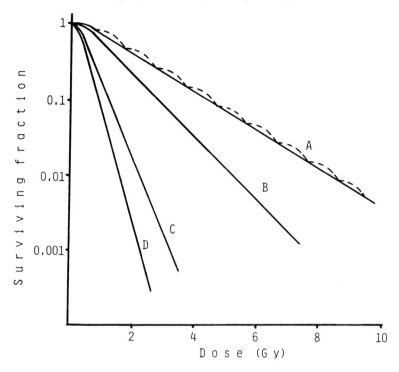

Figure 8.1. Survival curves of cells after acute or protracted doses of X-rays or neutrons
Curve A = X-rays @ $0\cdot01$ Gy min^{-1}.
Curve B = X-rays @ $1\cdot0$ Gy min^{-1}.
Curve C = neutrons @ $0\cdot01$ Gy min^{-1}.
Curve D = neutrons @ $1\cdot0$ Gy min^{-1}.

figure 8.1). The major dose rate effects occur between $0\cdot1$ Gy h^{-1} and 1 Gy min^{-1}. Dose rates less than $0\cdot1$ Gy h^{-1} generally produce no further sparing effect and dose rates in excess of 1 Gy min^{-1} are rarely more effective per unit dose than those at 1 Gy min^{-1}. Besides dose, dose rate, and dose fractionation the important factor of the linear energy transfer (LET) of the radiation can modify its biological effectiveness (Chapter 1). Protons, such as recoil protons produced in neutron irradiations (see Chapter 1), and α particles have high LET values and their tracks consist of dense columns of ionizations and excitations. X- and γ rays produce tracks of relatively sparsely ionizing electrons with low LET values.

It was mentioned in Chapter 1 that the LET of radiation has a marked influence on its relative biological effectiveness (RBE), and figure 1.12 showed a generalized curve of this relationship. Figure 8.2 shows the effect of LET on the survival of human kidney cells in culture. The first point to notice is that as the LET of the radiation increases, more and

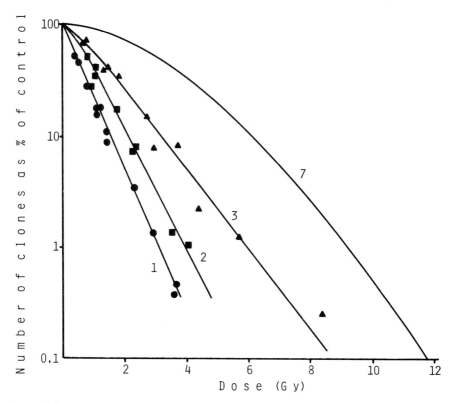

Figure 8.2.   Human kidney cell survival curves following radiation of varying linear energy transfers (LETs)
Source: G. W. Barendsen *et al.,* 1963, *Radiation Research* 18, 106; courtesy the authors and publisher.
Curve 7 = 2·5 keV μm^{-1} X-rays.
Curve 3 = 24·6 keV μm^{-1} α particles.
Curve 2 = 60·8 keV μm^{-1} α particles.
Curve 1 = 85·8 keV μm α particles.

more cells are killed for a given dose of radiation. The relative biological effectiveness of the radiation for cell killing increases with increasing LET. The biological effectiveness of one radiation relative to another is defined as the inverse ratio of the respective doses needed to bring about the same biological effect. This is illustrated in figure 8.3, which shows a generalized survival curve for X-rays, with its shoulder at low doses followed by the exponential portion and a similar survival curve for cells following neutron irradiation. The latter has a much reduced shoulder and is slightly steeper. The dose of X-rays (4 Gy) producing effect E_1 is twice the dose of neutrons (2 Gy) producing the same effect E_1. So the RBE of neutrons relative to X-rays for effect E_1 is 2 (4 Gy divided by 2 Gy). Figure 8.2 also illustrates the very important point that the RBE

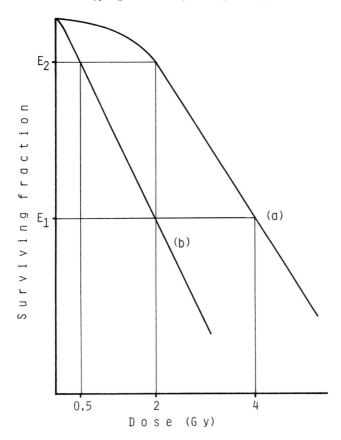

Figure 8.3. **Typical cell survival curves after (a) X-rays and (b) neutrons to illustrate the concept of relative biological effectiveness (RBE)**
At survival level E_1, neutron RBE is 2 (4 Gy/2 GY) and for survival level E_2, the neutron RBE is 4(2 Gy/0.5 Gy)

of neutrons increases as the dose decreases. So, for effect E_2 the ratio of the X-rays and neutron doses—2 Gy divided by 0·5 Gy—gives an RBE of 4 for neutrons. This tendency for RBE to rise quite spectacularly at lower doses is illustrated in figure 8.4 and this phenomenon has important consequences for radiation protection where low dose effects are particularly relevant, and for radiotherapy whenever multiple small doses of neutrons are involved.

Let us return to figure 8.2. Besides the increasing RBE with increasing LET of the α particles there is another aspect worth noting about the series of survival curves. At the low LET value of 2·5 keV μm^{-1} the X-ray survival curve has a large shoulder, which suggests that a multi-event process is the cause of cell death or effective repair is operative (see Chapter 3). At the very high α particle LET values, 85·8 keV μm^{-1}

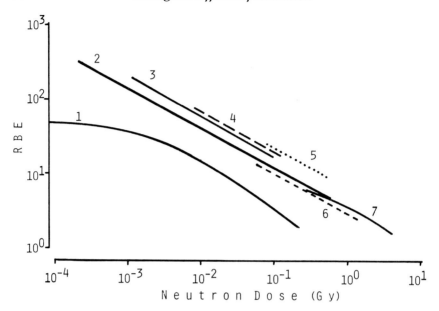

Figure 8.4. A plot of the logarithm of RBE versus the logarithm of neutron doses for a variety of biological effects
Source: H. H. Rossi, 1980, *Advances in Radiation Protection and Dosimetry in Medicine,* Plenum Press, p.138; courtesy the author and publisher.
Curve 1 = plant mutations.
Curve 2 = lens opacities.
Curve 3 = breast cancers.
Curve 4 = human leukaemia.
Curves 5 and 6 = chromosome aberrations.
Curve 7 = cell survival.

(curve 1), the shoulder has disappeared and the survival curve is purely exponential, which suggests that cell inactivation at these very high LET values is predominantly a single-hit phenomenon. At high LETs there is enough energy in the single high-energy track to kill cells, while at low LETs several sparsely ionizing tracks may have to co-operate in some way to cause cell killing.

Since the survival curves for high LET radiations are truly exponential and lack a shoulder, fractionation of the dose does not produce a reduced killing or 'sparing' effect. No recovery is possible with very high LET radiations.

8.3. Biological factors that influence the effect of radiation

Once again several of these have already been discussed in previous sections of this book. First, the importance of the phases of the cell cycle in relation to the efficiency of radiation cell killing was considered in detail in Chapter 4, as was the vital role of biological repair mechanisms in modifying the amount of biological damage. Second, the concept of the radiosensitivity and radioresistance of tissues was discussed in terms of their proliferative abilities and the law of Bergonie and Tribondeau (Chapter 6). Any change in the proliferative activity of a tissue can radically alter its response to a dose of radiation. Third, in Chapter 6 the modification of acute radiation injury in the gastrointestinal and bone marrow syndromes was discussed. Fourth, different species show differing susceptibilities to radiation, as seen from the table of $LD_{50/30}$ doses on p. 97. Finally, in any species, including man, genetic factors (see pp. 44 and 76) and hormonal balance can play a role in the alteration of radiosensitivity, although the reason for the differences are often far from understood.

8.4. Chemical factors that influence the effect of radiation

The chemical factors that influence the effect of radiation can be divided into two groups, the 'sensitizers' and the 'protectives'. Those factors which increase the effectiveness of a dose of radiation are the sensitizers, of which oxygen, hypoxic cell sensitizing drugs and the halogenated pyrimidines are the best known. In contrast, the factors which decrease the effect of a given dose of radiation are called protectives and include such componds as cysteine, cysteamine and glutathione, all of which contain the sulphydryl (–SH–) group.

The oxygen effect

In the presence of molecular oxygen (O_2) almost all biological systems so far tested are more sensitive to X- and γ rays than when they are irradiated at very low levels of oxygen (hypoxia) or in the absence of oxygen (anoxia). This ability of oxygen to enhance the effectiveness of a dose of radiation is known as the oxygen effect, and it is one of the most fundamental phenomena of radiobiology.

Oxygen modifies the quantitative amount of radiation damage, but does not alter it qualitatively; it merely reduces the dose of radiation required to give a certain radiobiological effect.

It is currently believed that the oxygen sensitization occurs via a fixation process. This involves oxygen combining with the free radicals formed in the target molecule and so producing peroxy-radicals (Chapter 1). This fixation of radiation damage occurs within 10^{-2}–10^{-3} μs of the irradiation. An alternative hypothesis, the electron transfer model, suggests that the radiosensitizing properties of oxygen and other radio-sensitizers are related to their electron-affinic properties. As we saw in Chapter 1 radiation can directly ionize target molecules and the free electrons so produced can either recombine at the initial site of their production ('self healing') or move along the molecule to an electron trapping site. The longer the electrons are free before recombination the more damage they may do. Electron affinic agents including oxygen may react with these free electrons and thus prevent recombination and so allow more damage to occur. Recent developments of rapid mixing of oxygen with cell suspensions have shown that the rise in the sensitizing effect is biphasic. The rapid component has been taken to indicate oxygen damage to outer cell membranes and the slower component may be due to damage to intranuclear sites (DNA?).

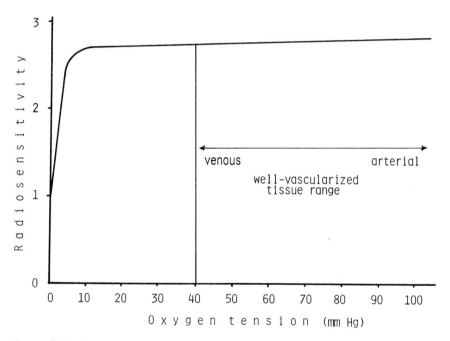

Figure 8.5. Curve for the relationship between relative radiosensitivity of a biological system and the oxygen tension
Source: L. H. Gray, 1957, *British Journal of Radiology* **30**, 406; courtesy the publisher.

Figure 8.5 shows the general relationship between the radiosensitivity of a biological entity and its oxygen tension at the time of irradiation. The graph shows the way in which the radiosensitivity varies very rapidly with oxygen tensions less than about 15 mm Hg, whereas increasing it above about 30 mm Hg leads to very little increase in radio-sensitivity. This detailed relationship between oxygen tension and radiosensitivity has been demonstrated for bacterial cell killing, mammalian cell killing and chromosome damage. The important point to note is that it is the oxygen tension inside the cells or tissues at the time of radiation that alters their radiosensitivity, and not merely the partial pressure of oxygen in the environment. Obviously an equilibrium is normally reached at which the intracellular oxygen concentration is approximately equal to the extracellular oxygen concentration. In mammals the oxygen tension of the arterial blood is about 70 mm Hg and of the venous blood about 40 mm Hg. The range of oxygen tensions of most well-vascularized tissues in mammals lies between these two extremes, so most normal tissues *in vivo* are maximally sensitive to radiation as far as the oxygen effect is concerned. This means that they are between two and three times as sensitive as they would be if no oxygen were present in the tissues and this factor of two to three is known as the oxygen enhancement ratio (OER).

Figure 8.6 shows the single dose survival curves for well oxygenated

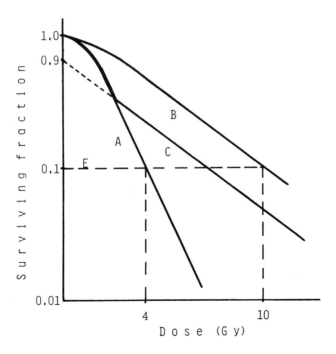

Figure 8.6. Survival curve for single doses of low LET radiation for populations of well oxygenated cells, A; poorly oxygenated cells, B; and a population containing 10% oxygenated and 90% (0.9) hypoxic cells, C
The OER at all levels of survival is 2·5, at the 0·1 level this is 12 Gy/5 Gy.
E = level of effect.

cells (curve A) and for very poorly oxygenated cells (curve B). It can be seen that for a level of effect E the dose under oxygenated conditions (4 Gy) is 2·5 times less than the dose required under hypoxic conditions (10 Gy). Since oxygen is truly a dose modifying agent the ratio of the slope of the two curves, i.e. their D_0s, will also be given an OER of 2·5. The survival curve C shows the bi-phasic response of a mixed population of hypoxic and well-oxygenated cells. The survival curve initially follows the oxic curve, A, and then deviates, becoming parallel with the hypoxic curve, B. The proportion of hypoxic cells in the population can be gauged by extrapolating the hypoxic part of the curve to the dose axis (the dashed line) and this gives a value of 0·1, i.e. 10 per cent hypoxic cells, 90 per cent oxygenated cells. Figure 5.1 shows an experimental example of such a discontinuous survival curve for cells obtained from multi-cell spheroids. There is currently much work on the varying oxygenation states to be found in spheroids but the working hypothesis is that the 'resistant tail' to the survival curve is most likely due to hypoxia in a percentage of the cells of the spheroid.

Figure 8.7 shows the oxygen effect *in vivo* for whole body doses of low LET radiation given to mice assayed for gastrointestinal death.

The statement that in the presence of oxygen all biological systems are between two and three times as sensitive to radiation as under anoxic

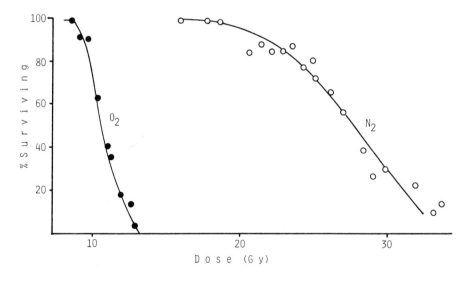

Figure 8.7. **The effect of oxygen on the percentage of mice surviving to 4 days after whole body irradiation with fast electrons of unanaesthetized animals breathing oxygen or nitrogen**
Source: S. Hornsey, 1967, in *Radiation Research,* edited by G. Silini, North-Holland Publishing Co. Ltd.; courtesy the author and publisher.
The OER is 2·7 at all levels of survival.

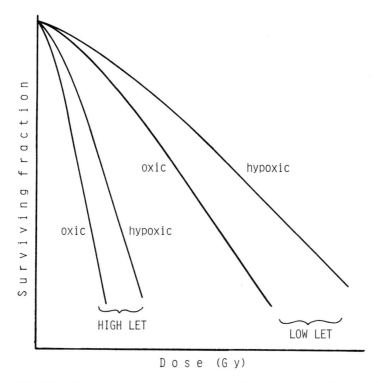

Figure 8.8.   **The effectiveness of oxygen as a modifying agent for high and low LET radiation**

conditions only holds true for relatively low LET radiations. As the LET rises, so the OER falls, and at very high LET values hypoxic and oxygenated cells are killed with almost equal facility, i.e. the OER tends to unity.

Figure 8.8 shows pairs of survival curves for cells irradiated with high and low LET radiations under oxygenated and hypoxic conditions. It shows that the OER for the former is close to one, i.e. oxygen does not really modify the effect of high LET radiation.

Figure 8.9 shows the general relationship between OER and LET with the OER reaching a value close to one for mammalian cells at LETs between about 100 and 200 keV μm^{-1}. But why should there be a reduction in the oxygen effect with an increase in LET? The most likely explanation, and one for which there is some experimental evidence, is that oxygen is actually produced in the dense ionization tracks of high LET radiations and this actually oxygenates and so sensitizes the cells at the instant of the irradiation. Superimposed on this curve is the RBE versus LET curve taken from figure 1.12 showing its peak at about 110 keV μm^{-1}.

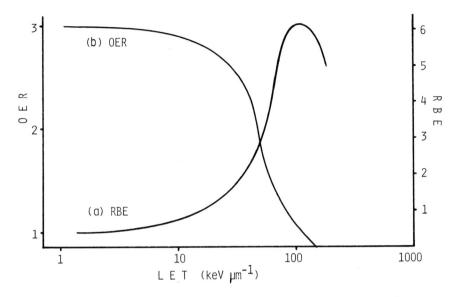

Figure 8.9. The relationship between the linear energy transfer of radiation (LET) and (a) its relative biological effectiveness (RBE); (b) the oxygen enhancement ratio (OER)

Hypoxic cell sensitizers

Apart from oxygen the most important group of sensitizing agents are the chemical compounds known as 'hypoxic cell sensitizers'. As their name implies these drugs are capable of selectively increasing the sensitivity of hypoxic cells to the lethal effects of radiation. They appear to act like oxygen, and like oxygen they have a strong affinity for electrons. They do not sensitize well oxygenated tissues, i.e. normal tissues, and they are capable of diffusing much further into tissues than oxygen. Sensitizers are to be found among such compounds as quinones, acetophenones, nitrofurans, glyoxals and the nitroimidazoles. The most promising drug so far is 'misonidazole' (a nitroimidazole) which has been shown to enhance the radiosensitivity of hypoxic tumours cells in experimental animals by a factor of 2·5 and is currently undergoing clinical trials. The relevance of hypoxic cell sensitizers is discussed in Chapter 9.

Chemical radiation protective agents

The addition of chemical protective agents reduces the effectiveness of radiation by a factor of between 1·5 and 2·0. To be effective all protective agents need to be present at the time of radiation and they must be close to the critical site of radiation damage. Post-irradiation treatment with protectives is ineffective.

The principal group of protective agents, the sulphur-containing aminothiols and their disulphides, includes cysteine, cystamine, cysteamine, mercaptoethylguanidine (MEG), S-(2-amino ethyl) iso-thiuronium bromide hydrobromide (AET) and glutathione (GSH).

The most favoured hypothesis to explain the molecular mechanism of cellular radioprotection is that H atoms are transferred from the sulphydryl compound (M-SH) to the biological free radical (R'). The R' is thus repaired and converted to RH.

$$M\text{-}SH + R^{\bullet} \quad \rightarrow \quad RH + MS^{\bullet}$$
$$\text{repair}$$

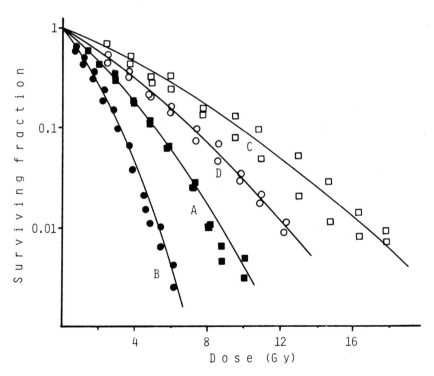

Figure 8.10.   X-ray survival curves for BUdR labelled and unlabelled cells in the presence and absence of IM dimethyl sulphoxide (DMSO)
Source: J. D. Chapman, 1980, in *Radiation Biology in Cancer Research*, Raven Press; courtesy the publisher.
Curve A – unlabelled cells.
Curve B – BUdR labelled cells.
Curve C – unlabelled cells plus DMSO.
Curve D – BUdR labelled cells plus DMSO.

There is direct experimental evidence using pulse radiolysis of the repair of simple free radicals by hydrogen transfer from sulphydryl compounds.

A nice example of both radiosensitization and radio-protection is given in figure 8.10. This shows survival curves for X-irradiated mammalian cells that are BUdR-labelled (curve A) and BUdR-unlabelled (curve B). The labelled cells are more radiosensitive than the unlabelled cells and this has been attributed to the increased damage in the DNA of the BUdR-labelled cells (see Chapter 3, page 62). Curves C and D show the effect of the presence of a radioprotective agent, dimethyl sulfoxide (DMSO). This agent acts by scavenging OH• radicals that are responsible for the indirect action of radiation. Figure 8.10 shows that DMSO is able to protect both BUdR-labelled and -unlabelled cells.

8.5. Summary

The effect of a dose of radiation on a biological entity can be influenced by physical, biological and chemical factors.

Among the physical factors are the absolute size of the dose, the rate at which the radiation is given, whether the dose is single or fractionated and the linear energy transfer of the radiation.

The important biological factors that modify the final expression of radiobiological damage include the proliferative state of the cells, the phase of the cell cycle at irradiation, the presence of biological repair capacity and sex, age and species differences.

The chemical factors that influence radiation damage can be divided into sensitizers (e.g. oxygen and the hypoxic cell sensitizers) and protectives (e.g. cysteine, glutathione and other sulphydryl-containing compounds).

Chapter 9
Radiation and cancer

9.1. Introduction

We saw in Chapter 6 that radiation damage to the central nervous system, the gastrointestinal system, and the bone marrow can be lethal. In these syndromes we saw that the nature of the radiation injury could be pinpointed with some accuracy and the course of the lethal effects could be followed. However, besides these acutely lethal effects of radiation there are more subtle, insidious forms of injury known as the late or delayed effects, which do not appear until long after the irradiation. The principal late effects are genetic damage, the induction of cancer, the induction of cataracts of the lens of the eye and late radiation fibrosis and blood vessel damage. The most significant of the late effects is radiation carcinogenesis. However, as we shall see there are several reasons why it is difficult to assess the size of the risk. This statement is true even for moderate doses of 1–5 Gy. It is probably true to say that it will always be impossible to accurately assess the cancer risk at the small doses received by radiation workers or by the general public from medical or environmental exposure (see Chapter 11). As we shall see, the reasons for the uncertainty, and therefore the controversy, over cancer risks involve ignorance of how radiation induces cancer and how to interpret the existing animal and human data. Before tackling these problems, it will be useful to outline some basic facts about cancer.

9.2. Some facts and theories about cancer

A cancer (or tumour) is a group of cells that divides to give a rather loosely organized mass of cells. In the special case of leukaemia, cancer of the bone marrow and lymphoid tissues, the cells are not aggregated

147

together and some of them are free to circulate in the blood and lymph vessels, while others are bound to different tissues. Tumour cells do not generally show any special physiological or tissue function, although some produce physiologically active substances such as hormones. The cells of many tumours seem to have surface properties that allow them to break away from neighbouring cells and infiltrate and invade other tissues. Such cells may form secondary tumours (metastases) in a new site if conditions for growth are favourable. Cancer cells are also characterized by the fact that they are relatively independent of the normal control mechanisms of the body. An adequate definition of cancer would have to include these three attributes: proliferative ability, metastatic capacity, and a relative autonomy from normal homeostatic mechanisms of the body.

Physical, chemical and viral agents can cause normal cells to become cancerous. These cancer-forming agents or carcinogens include ionizing radiation, ultraviolet radiation, thousands of organic chemicals, some hormones and many viruses. One estimate puts the proportion of human cancers induced by environmental agents as high as 80 per cent. Genetic make-up can also predispose an animal population to a very high incidence of 'spontaneous' tumours. Inheritance also plays a role in human cancer proneness. The tumours caused by all the different carcinogens are not really distinguishable, but this similarity does not mean that all the different agents cause cancer by the same mechanism; there may only be superficial resemblance and further work may reveal detailed differences.

What are the mechanisms that cause normal cells to become cancerous? At present there is no real answer and it seems true to say that until there is further understanding of the control mechanisms of normal cell division, there is little chance of a better understanding of the cancer cell, which is essentially a cell dividing out of turn. Despite our basic ignorance, there is no lack of theories as to how radiation (or other carcinogens) might induce cancer. Most theories concentrate on the fact that the alteration of a normal cell involves its inability to integrate with its neighbouring cells. At the same time the alteration does not involve any interference with the cell's proliferative ability, which is usually enhanced.

The somatic mutation theory and the viral theory of radiation carcinogenesis have received the most support.

The somatic mutation theory of cancer suggests that the DNA of a cell becomes altered or mutated so that its information content is changed. The mutation might take the form of an invisible gene mutation, or an actual breakage or loss of a chromosome might occur. If such a defective cell is capable of division independently of the control mechanisms of the body, it is tantamount to a cancerous cell. These somatic mutation theories are so vague that they amount to little more than a re-description of the term 'cancer', for example, a cancer cell is a cell with a mutation giving it cancerous potentialities. Nevertheless, the theory is an attractive one for several reasons.

First, a number of bacterial and mammalian cell systems have been developed to identify and screen potentially hazardous chemicals for their mutagenic activity. These studies have led to the conclusion that the overwhelming majority (over 90 per cent) of chemical carcinogens cause cancer by damaging DNA, i.e. causing a somatic mutation. Radiation is of course a potent mutagen (see Chapter 7) and, as we shall see, a carcinogen as well.

Second, as we have mentioned elsewhere in this book there are certain rare human diseases (Xeroderma pigmentosum, Fanconi's anaemia, Bloom's syndrome, Ataxia telangiectasia and others) in which the patients are predisposed to develop cancer, show instability in their chromosomes and varying degrees of deficiency for DNA repair capacity. Many workers feel that this spectrum of diseases provides the best evidence for a causal link between cancer and damage to gene structure and/or function in somatic cells of the body.

Third, modern cytogenetic methods have shown that almost all tumours in animals and man have chromosome abnormalities and in many cases the same abnormality appears in all the cells of a tumour, indicating that it may have originated from a single cell. However, the wide variety of aberrations to be found in tumour cells leads many cancer biologists to conclude that the aberrations are a consequence of cancer growth rather than its cause. In contrast to the lack of consistency of the aberrations in many cancers there are indeed consistently recurring abnormal chromosome arrangements. The best known of the latter is the famous 'Philadelphia' chromosome (Ph) found in over 95 per cent of patients with chronic granulocytic leukaemia. Overall, the chromosome study of cancer is still equivocal and it would be wrong to state that aberrations are a *sine qua non* for cancer.

Fourth, there is some intriguing evidence linking malignancy with DNA coming from recent experiments involving the fusion (hybridization) of malignant and non-malignant cells. The work suggests that all cells are potentially cancerous, i.e. carry a malignancy factor (a set of genes?) that is recessive and is therefore normally suppressed by another set of genes for non-malignancy. So a cancer cell is a cell that has lost the genes that control malignancy. Fusion of normal cells with malignant cells produces hybrid cells that are non-malignant because the controlling set of genes from the normal cells has suppressed the malignancy factor of the malignant cells. These non-cancerous hybrid cells can become malignant when they lose certain chromosomes after undergoing cell division.

For these and other reasons the somatic mutation theory is currently the most favoured mechanism to explain cancer induction. Radiation damages DNA in many ways (see Chapter 2) and it may be that DNA mutations in somatic cells are the cause of cancer.

Many viruses are known to be associated with cancer cells and with an increased incidence of cancer, and it has been suggested that radiation might stimulate a dormant carcinogenic virus. Alternatively, radiation is known to suppress the immune response and this might leave an animal open to attack by a carcinogenic virus.

Both the somatic mutation and viral theories involve intra-cellular changes, but no one has excluded the possibility that damage outside the cells may promote cancer. Radiation might disorganize the extracellular matrix in such a way as to allow the appearance of cancerous growth.

9.3. Induction of cancer by radiation

One of the first observations of cancer following radiation was the appearance of skin tumours on the hands of many of the early workers with X-rays. Since that time many systematic studies on animals have shown that radiation causes an increase in the incidence of almost all types of naturally ('spontaneously') occurring cancers. The malignant tumours do not appear until long after the exposure, the delay may be as long as 30–40 years in the case of some human cancers. Between the exposure and the appearance of the tumour there may be no observable defect in the tissues that eventually become cancerous.

Radiation is capable of inducing tumours in almost all of the tissues of the body, although tissues vary greatly in their susceptibility to radiation-induced cancer. Generally speaking tissues that have a high rate of cell division are more prone to tumour induction than tissues that exhibit a low rate of cell proliferation. The fact that radiation may induce tumours in nearly all tissues distinguishes it from the vast majority of chemical and viral agents that can cause tumours only in a few selected tissues.

As is the case for many other effects of radiation, a high dose rate is generally more effective in causing tumours than a low one, and high LET radiation is generally more effective than sparsely ionizing radiation in causing tumours.

Tumours most commonly appear in the tissues that have been directly exposed to radiation, but there are some cases where cancer induction is an indirect process. For example, if mice with their thymus glands removed are irradiated and a new thymus is then transplanted into them a cancer will develop in this unirradiated organ. Or again, the pituitary gland may become cancerous as a result of the irradiation of the thyroid gland with radioactive iodine (^{131}I).

We must return and examine in more detail some of the general statements made in the introduction concerning the uncertainties and controversy that surround this subject of radiation carcinogenesis. We shall divide the discussion into three parts, essentially the three sources of information that are available for estimation of cancer risk rates. These are: (1) certain aspects of the fundamental radiobiology of dose response curves, (2) experimental animal data and (3) the human epidemiological data.

9.4. Fundamental aspects of dose response curves for cancer induction

In this section we shall consider the question "what is the shape of the curve relating the incidence of cancer with the radiation dose?" It should be stated at the outset that this question will not be satisfactorily

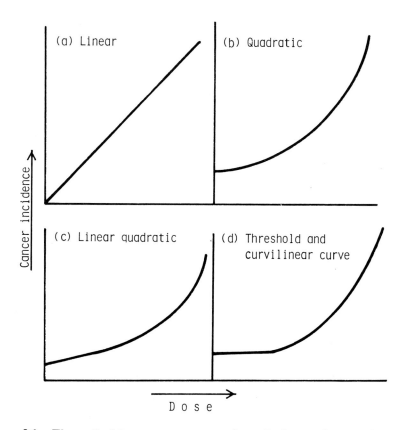

Figure 9.1. Theoretical dose response curves for radiation carcinogenesis

answered until there is a greater understanding of the mechanism of carcinogenesis. Nevertheless, certain assumptions about the shape of the dose response curve have to be made and can be made using basic radio-biological concepts. These theoretical curves can then be compared with the experimental data from animal studies and with the human epidemiological data to see which, if any, of the general shapes is consistent with the empirical evidence.

Figure 9.1 shows a variety of theoretical dose response curves for radiation-induced cancer.

The linear relationship between dose and incidence of induced cancer was first adopted some 25 years ago as the simplest working hypothesis for radiation protection purposes. It is still held by many authorities to be consistent with much of the experimental and human data. The linear function is given by

$$I = c + aD$$

where I is the cancer incidence after dose D, c is the cancer incidence of the unirradiated, control population, and a is the coefficient that determines the slope of the linear curve (figure 9.1a).

If two independent initiating events are required to transform a normal cell into a cancer cell then the form of the dose response curve would be a quadratic given by the function

$$I = c + bD^2$$

in which the cancer incidence (I) varies with the (dose)^2 (figure 9.1b).

A somewhat intermediate dose response relationship is the linear-quadratic (LQ) form expressed by the function

$$I = c + aD + bD^2$$

which combines the first two equations. This LQ model has received much attention in the last few years because certain of the principles of microdosimetry, which are outside the scope of this book, would favour such a hypothesis. The important point about the LQ model is that the initial part of the dose response curve, i.e. the low dose region, will show no threshold and will be linear. At higher doses the curve will increase more steeply than linearly (figure 9.1c).

Finally, figure 9.1d shows a theoretical curvilinear relationship which has a threshold of dose, i.e. a dose at which there is no risk of cancer induction. We need not concern outselves with figure 9.1d since it is not possible to prove experimentally the existence of a threshold at least at low doses.

As we shall see both human and experimental cancer induction curves often tend to plateau or even show a peak incidence at a certain 'turnover dose point' after which the cancer incidence falls with increasing dose.

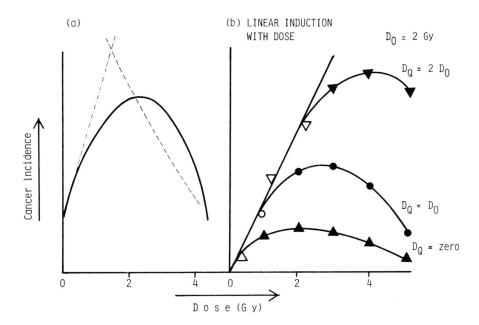

Figure 9.2. (*a*) **The observed cancer incidence (―――――) as a function of the completion between induction (- · - · - · - · -) and cell sterilization (- - - - -)** (*b*) **Diagrammatic representation of the way in which the shape of the cell killing response modifies the observed cancer yield**
Source: R. H. Mole, 1975, *British Journal of Radiology* **48**, 157; courtesy the author and publisher.
Induction is linearly proportional to dose. The open symbols are points unaffected by cell killing, the black symbols are points where the observed yield is reduced in proportion to cell sterilization.

Figure 9.2 show a theoretical relationship of this type. The seemingly complex curve is explained by assuming a competition between two processes, (1) the induction by radiation of a malignant transformation of a normal cell and (2) the possibility that such a transformed cancer cell will be killed or at least lose the ability to divide. In Chapter 3 we saw that cell inactivation is exponential with dose and may be represented by the equation

$$f = e^{-(\alpha D + \beta D^2)}$$

And if we combine this equation with the LQ equation we get the most general function for the dose response relationship

$$I = (c + aD + bD^2)e^{-(\alpha D + \beta D^2)}$$

which with suitable coefficients will fit the shape of the curve in figure 9.2*a*.

Figure 9.2*b* illustrates the effect that changing the parameters of the cell inactivation term would have on the final cancer incidence curve. For example, if there were no shoulder to the killing curve ($D_q=0$) the initial linear part of the induction curve rapidly plateaus and tends to decrease as the dose increases.

As we shall see in the next sections both the human and the experimental cancer induction curves show a wide variety of shapes. We shall also see that the individual data points that make up a cancer induction curve usually have rather large confidence intervals. In fact, such is the variability of the data that statistical curve fitting is often inadequate to distinguish between the different models of dose response relationships in figure 9.1. This is especially true at the low dose region of the curves, i.e. the region of greatest interest for the assessment of cancer risk in radiological protection.

All but one of the major national and international committees that considered cancer risk estimation in the 1970s favoured a linear no-threshold relationship for human radiogenic cancer. The committees all

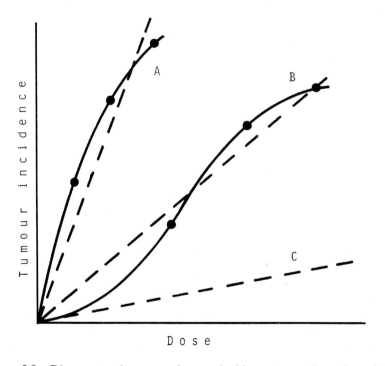

Figure 9.3. Diagrammatic curves of cancer incidence versus dose after a high LET (curve A) and after a low LET curve B). Curve C shows the likely reduced effectiveness of low dose rate, low LET radiation
After R. J. M. Fry, 1981, *Radiation Research* 87, 224; courtesy the author and Academic Press.

accepted that a number of non-linear curves could be drawn through the limits of error on the data. They were simply affirming that the linear fit was not only the simplest and administratively convenient but was also 'consistent with the data'. The linear hypothesis also has the merit that it is conservative. For low LET radiation it may overestimate the real cancer risk at low doses. Curve B in figure 9.3 is an idealized curvilinear cancer induction curve for low LET radiation. The dotted line drawn through the curve to the origin is the linear extrapolation that would be used to estimate the cancer incidence at the low dose region where there are no dose points. One can see this linear extrapolation will overestimate the risk in the low dose region. In contrast, for high LET radiation, since in some experiments the data points turn over at relatively low doses (curve A), linear extrapolation consistent with the points might underestimate the low dose cancer risk for high LET radiation.

Figure 9.3 also illustrates another important point—the influence radiation quality has on the cancer incidence curve. As with other radiobiological endpoints, high LET radiation is more effective than low LET radiation (see figures 8.2 and 8.3) and because of the shape of the curves the RBE values are likely to rise sharply at low doses.

Another factor that often modifies the efficiency with which radiation induces cancer is dose rate. This will be dealt with in detail in the next section.

9.5. Experimental radiation carcinogenesis

The induction of fatal cancer is regarded as the most significant risk of radiation to man. Consequently much effort has been put into experiments to understand the mechanisms of cancer induction and to provide answers to such important questions as: What is the shape of the dose response curve at low doses? Does the shape vary from tissue to tissue? What effect do radiation quality and radiation dose rate have on the dose response curve for cancer? And finally, to what extent are experiment dose effect relationships, obtained using animals, relevant to human radiogenic cancer risk rates?

Figure 9.4 shows a small selection of the dose response curves that have been obtained in mice and rats for a variety of tumours. The immediate impression is the wide range of shapes, from linear to highly curved, some with plateaux, some with turnover points (see figure 9.2a), some with thresholds, and so on. This variability is not surprising when one considers the multiplicity of factors that may intervene between the initiation of a cancer cell and the final appearance of a cancer. Immunological, hormonal, vascular and nutritional factors together with cell proliferation kinetics may all influence the final cancer incidence, and

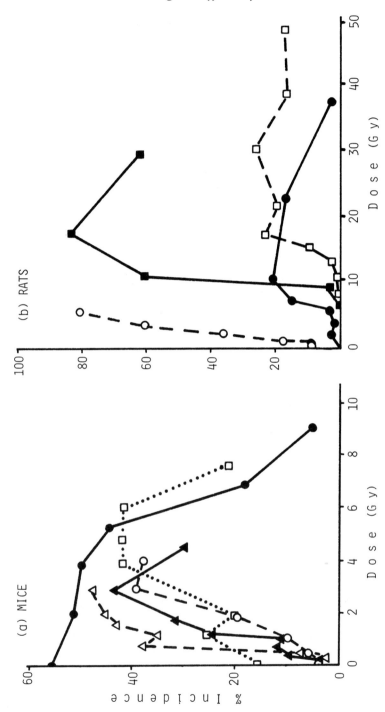

Figure 9.4. (*a*) Dose effect relationship in mice for the induction of myeloid leukaemia (▲), thymic lymphoma (○), reticular tumours (●), ovarian tumours (△) and lung tumours (□). (*b*) Dose effect relationship in rats for the induction of mammary tumours (○), kidney tumours (■) and skin tumours (□, ●)

we saw in figure 9.2*b* how small alterations in the cell-killing parameters (D_o, D_q) drastically alter the final shape of the induction curve.

One of the important lessons from experimental radiation carcinogenesis is that the variety of dose response curves probably implies different mechanisms for different types of tumours.

The variables above contribute to the large standard deviations that are a characteristic of the individual points on many cancer incidence curves (see figure 9.7). Such statistical variations usually make it impossible to define accurately the exact shape of the cancer incidence at low doses ($\langle \langle 0.5$ Gy), i.e. the region of greatest interest to radiological protection. This is true even for the largest and most rigorously controlled experiments.

The statistical problems of the low dose region also mean that experimental carcinogenic data cannot be used, with any confidence, to provide numerically useful data for human risk estimation. The uncertainties of extrapolating mouse data to man outweigh the uncertainties and flaws that characterize the human data itself (see section 9.6).

In vitro assay systems have recently been developed in which it is possible to quantify the number of normal cells that have been transformed by radiation into malignant cells. At present the technique only works with a very few cell types. It involves irradiating a known number of cells on a petri dish and then observing the subsequent colony growth. The colonies produced by transformed cells are very distinctive and the cells of such colonies can be tested *in vivo* for their malignancy. Less than one cell in a million will spontaneously transform into a cancer cell. Using these systems it is possible to obtain rather precise dose response relationships over a wide range of doses, starting at dose levels that are impractical in animal experiments, for example 0.01 Gy or below. These transformation incidence curves are proving to have unexpectedly complex shapes. Figure 9.5 shows such a curve for single doses of X-rays. It seems to consist of three parts. At doses above 1 Gy the data are consistent with a quadratic dependence on dose. Below 0.3 Gy the data show that the number of transformed cells is directly proportional to the dose, i.e. a linear relationship, while between 0.3 and 1 Gy the frequency of transformed cells does not vary with dose. The data as a whole cannot be fitted by a single straight line and a linear regression line—the dotted line in the figure—would underestimate the incidence of transformed cells at low doses. These complex *in vitro* curves are as yet inexplicable. One might have expected a somewhat simpler dose response relationship *in vitro* since the systems assay cancer initiation and do not involve any of the complicating *in vivo* factors noted earlier.

So the problem of the exact shape of the low dose region of cancer

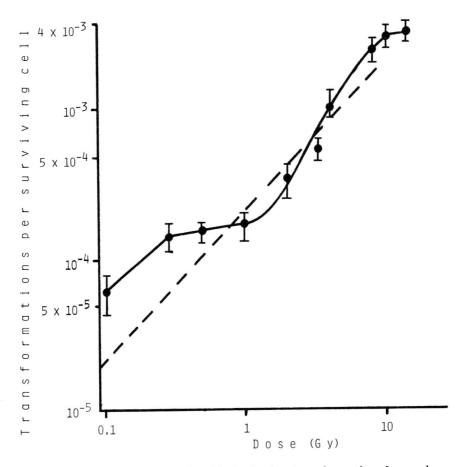

Figure 9.5.   Dose-response relationship for *in vitro* transformation of normal cells to malignant cells following exposure to single doses of X-rays
Source: E. J. Hall and R. C. Miller, 1981, *Radiation Research* **87**, 208–223; courtesy the authors and publisher.

induction curves remains, but may one day be clarified, if not solved, by more and better animal experiments.

In Chapter 8 we saw that protracted or fractionated doses of low LET radiation are generally less biologically effective than single acute doses. This is also true for cancer induction. The discussion of what factor ought to be introduced to allow for the possible reduced effectiveness of low dose rate in carcinogenesis is always a topic of lively debate in radiation protection circles. Since there is so little useful human data more or less regard has to be paid to animal evidence. Curve C in figure 9.3 shows the theoretical cancer incidence expected for low LET radiation at a low dose rate. It is derived by linearly extrapolating the low dose region of curve

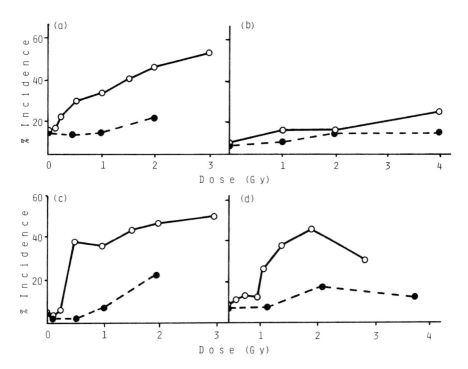

Figure 9.6. The influence of dose rate for a variety of experimental mouse tumours. (*a*) **Thymic lymphoma at either 0·45 Gy min⁻¹ (○) or 0·08 Gy min⁻¹ (●); (*b*) mammary tumours at either 0·45 Gy min⁻¹ (○) or 0·08 Gy min⁻¹ (●); (*c*) ovarian tumours at either 0·45 Gy min⁻¹ (○) or 0·08 Gy min⁻¹ (●); (*d*) myeloid leukaemia at either 0·8 Gy min⁻¹ (○) or 0·75 Gy day⁻¹ (●)**
Source: (*a*), (*b*) and (*c*) R. L. Ullrich and J. B. Storer, (*d*), A. C. Upton, 1980, *National Council on Radiation Protection and Measurements Report*, No. 64; courtesy the authors and publisher.

B. Figure 9.6 shows some experimental animal data for cancer induction at high and low dose rates. A 'dose rate effectiveness factor' (DREF) can be derived from such curves by taking the ratio of the slopes of the high and the low dose rate curves. Values between 1 and 10 have been obtained for such DREFs for different tumours induced, i.e. for some tissues changes in dose rate do not alter the number of tumours, while for other tissues radiation at a high dose rate is ten times as effective as low dose rate radiation. Recognition of such dose rate effects in radiation protection is important since human occupational exposure is usually at very low dose rates while cancer risk estimates are generally derived from acute, high dose rate exposure data. Incidentally, there are some animal data to indicate that for high LET radiation (fission neutrons, etc.) protracted and fractionated doses are more effective at inducing tumours than acute doses. At present there is no entirely satisfactory explanation for such findings.

Finally, there is considerable experimental evidence that high LET radiation is more effective at inducing cancer than low LET radiation. A wide variety of RBE values (see Chapter 8) can be found in the research literature. RBE values at doses greater than 1 Gy are often close to unity. However, because of the relative shapes of the high and low LET dose response curves (figure 9.3) RBE values tend to increase sharply and may be much greater than 10 for low doses (about 0·01 Gy) and low dose rate high LET radiation.

We have just touched upon some of the variables known to affect the dose response curve for radiation carcinogenesis. Table 9.1 lists some other factors which, if they were fully understood, would probably give us a fairly definitive understanding of radiation carcinogenesis.

Table 9.1. Factors which affect dose response curves for radiation carcinogenesis

Physical	*Biological*
Whole body, partial body	Age at exposure
External, internal radiation	Tissue at risk
Uniform or point source radiation	Sex
Single, fractionated, protracted doses	Hormonal status
Dose rate, radiation quality	Immunological status
	Genetic traits
	'Spontaneous' incidence
	Repair capacity
	Interaction with other chemical, physical or biological (viral, etc.) carcinogenenic agents

9.6. Human data

Radiogenic cancer in man has been studied since 1902 when a skin cancer was reported on the hand of a radiation worker. This was some seven years after Röntgen's discovery of X-rays. There can of course be no cancer induction experiments in man comparable with those in animals. Nevertheless, observations on people deliberately or accidentally exposed to radiation have shown that radiation doses too small to cause macroscopic tissue damage do carry an increased risk of cancer. It is widely accepted that this increased risk is approximately proportional to dose down to doses of the order of 0·01 Gy. The risk also applies to most of the organs or tissues of the body.

There are many problems associated with risk estimation:

1. The latent period between exposure and the appearance of a tumour may be decades, which necessitates long follow-up periods.

2. Dose assessment is often imprecise. For the A-bomb survivors the

dosimetry is all retrospective and is currently being revised, yet again (see below). For internal radiation there are problems of dose distribution and the choice of the relevant dose, especially where the radioactive substance has a long biological half-life. There is also often a chemical toxicity associated with the isotope which has to be taken into account; for example 'thorotrast' (thorium oxide) in the liver causes a great deal of cell death from chemical toxicity, and separating the chemical and radiation effects in assessing the risk of induction of liver cancer is impossible.

3. The dose levels may not be suitable for risk estimation. They may cover such a narrow range that linear extrapolation to different doses may be invalid (see figure 9.3). The dose distribution may also be important in risk assessment. It may not be justifiable to assume that a high dose to a small fraction of the body carries the same carcinogenic risk as the same total dose given uniformly to the whole body.

4. Which control populations are to be compared with the exposed population? This is particularly a problem with radiotherapy patients and studies may lead to false conclusions unless the correct control population is considered. For instance, it was thought that treatment with radioactive iodine for hyper-thyroidism increased the incidence of leukaemia, but when an equivalent group of people with hyper-thyroidism who had not been treated with radiation was studied it was found that these people had a high natural incidence of leukaemia. The selection of a control population to compare with the survivors of Hiroshima and Nagasaki is also difficult. A comparison with Japan as a whole, a largely rural population, is invalid because disease incidences are different in urban and rural communities, so a suitable urban Japanese population must be found. The control group most often used is the population in the two cities who received less than 0·1 Gy. But even then the exposed group is an unrepresentative urban population as there were no men of military age in the cities at the time of the bombings. Also, there was a great deal of immediate mortality which must have been to some extent selective and which may account for the generally lower risk estimates obtained from the A-bomb survivors compared with other irradiated groups.

5. Finally, there are statistical problems associated with the need to observe large populations for long periods to detect the small increases in tumour incidence. These are often insurmountable and lead to wide confidence limits on risk estimates. For example, if a population of 40 000 radiation workers received a dose of 1 cSv, then it can be shown that they would have to be studied for 30 years to detect a cancer induction risk of 10^{-4} cSv^{-1} (see also Chapter 11).

Despite these difficulties, there are more than a dozen different organs of the body for which there is good retrospective epidemiological data.

This involves populations of known size whose doses are well documented and who have been followed for prolonged periods and with an adequate control population. These populations are listed in table 9.2. Each population gives us data on different modes of exposure, whether as single, acute, whole body doses of external radiation (A-bomb survivors); partial body irradiation (radiotherapy exposures); or protracted, internal exposures (luminous-dial painters, uranium miners and thorotrast patients).

Table 9.2. Populations which can be used for cancer risk estimates

Nuclear Weapons
  A-Bomb survivors (Hiroshima and Nagasaki)
  Tests: Marshall Islanders
         Other fallout
Medical Radiotherapy
  Spondylitics
  Other non-malignant conditions, e.g., pelvis, scalp
Medical Diagnosis
  Fluoroscopy (TB)
  Thorotrast
  Pelvimetry (foetus)
Occupational
  Uranium and other miners
  Dial painters
  Radiologists
  Nuclear industry
Natural Environment
  Areas of high natural background

Thyroid, breast, lung and bone cancer and leukaemia are the best documented of human radiogenic cancers. Evidence of these five cancers comes from more than one source. In table 9.3 each + represents an independent epidemiological survey. Such independent risk estimates are of great corroborative value. Risk estimates for other cancers are less reliable but we shall return to the specific values of risk estimates later.

Table 9.3. Sources of risk estimates for different radiogenic cancers

	Leukaemia	*Thyroid*	*Breast*	*Lung*	*Bone*
Japaneses A-bomb survivors	+	+	+	+	–
Marshall Islanders	–	+	–	–	–
Radiotherapy patients	+++	++	+	+	+
Medical diagnosis	–	–	+	–	+
Uranium miners	–	–	–	+	–
Dial painters	–	–	–	–	+

Until 1980, the most useful source of information on the carcinogenic hazard of both high LET (neutrons) and low LET (γ rays) radiation came from the cancer incidences in the Japanese survivors of the atomic bombings. A joint Japanese–US team has followed the medical history of some 100 000 A-bomb survivors for over 30 years. However, a re-evaluation in 1980 of the dosimetry of the A-bombs showed significant differences from the previously computed dose values in 1965. It is generally agreed that a great deal of work needs to be done before the 'new' A-bomb dosimetry can provide a firm basis for a revised set of radiation risk standards. Figure 9.7 shows a preliminary assessment of

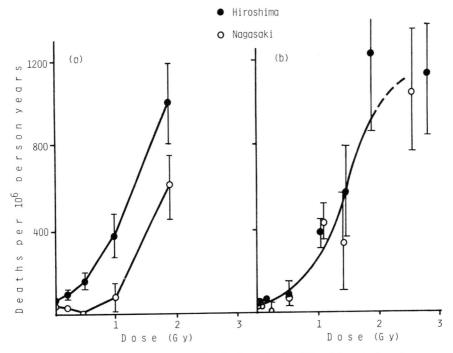

Figure 9.7. **Dose response curves for leukaemia for Hiroshima (●) and Nagasaki (○): (a) using the 1965 dosimetry level; (b) using the 1980/81 dosimetry level**
(a) *Source*: H. H. Rossi and C. W. Mays, 1978, *Health Physics*, **34**, 353–360; courtesy the authors and Health Physics Society. (b) *Source:* T. Straume and R. L. Dobson, 1981, Lawrence Livermore Laboratory Reprint, UCRL 85446; courtesy the authors.

the difference in the dose response curves for Hiroshima and Nagasaki using the old 1965 dose estimates (a) and the recent 1980 dose estimates (b). The difference between the cities in (a) was attributed to the high neutron component at Hiroshima. We have seen that neutrons have a higher relative biological effectiveness than γ rays. The 1980 re-evaluation of the A-bomb dosimetry suggests that there were virtually no neutrons at either city and that the γ dose was underestimated at Hiroshima and overestimated at Nagasaki. These adjustments effectively abolish the difference between the cities and both sets of data fit the same linear quadratic curve, at least up to about 2 Gy (see figure 9.7*b*).

Figure 9.7*b* also illustrates two important points already made in section 9.4. First, the wide confidence limits on the individual leukaemia incidence points make any risk estimates statistically uncertain. Second, the LQ dose response relationship means that the use of a linear extrapolation would tend to overestimate the real leukaemia risk at least for this set of data (see also figure 9.3).

In contrast to the LQ relationship in figure 9.7*b*, the bone cancer incidence in figure 9.8 shows a negative (upward) sloping relationship

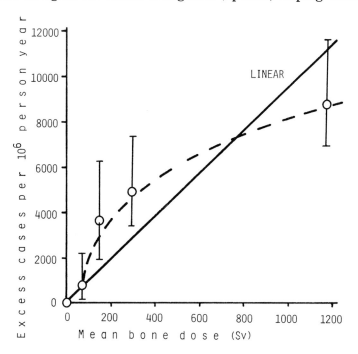

Figure 9.8. Dose response data for bone cancer in a group of radium-226 exposed dial painters and radiotherapeutically exposed patients
Source: National Research Council, Advisory Committee on the Biological Effects of Ionizing Radiation, 1972, *BEIR I Report;* courtesy the National Academy of Sciences, U.S.A.

with dose. The data come from persons exposed to radium-226 α particles between 1915 and 1935. The group includes dial painters and some patients given radium therapeutically. A linear extrapolation would

underestimate the real bone cancer risk for such a curve. These data and
those from other bone cancer studies show a tendency to plateau at high
doses.

The effect of dose and cancer induction was discussed in section 9.5 and
it was noted that, in general, it is expected that for low LET radiation the
risk per unit dose will decrease as the dose rate decreases. However, this
is not always the case and for cancer of the breast there is no significant
difference in the risk rate between women receiving several small doses in
multiple fluoroscopic examinations carried out over several years and
women receiving large doses from high dose rate radiotherapy. For high
LET radiation there is even data to suggest that protracted doses are
more effective than acute doses. For example, figure 9.9 shows the
incidence of bone cancers (osteosarcomas) in juveniles and adults after
intravenous injection of a colloidal preparation of the α-emitting isotope
radium-224 called 'peteosthor'. As the dose is spread out in time the
carcinogenic effect per unit dose increases from 40 per 10^4 per Gy to 200
per 10^4 per Gy of alphas. Such data are very relevant to any assessment
of α particle effectiveness when received over long periods, as in
occupational exposure.

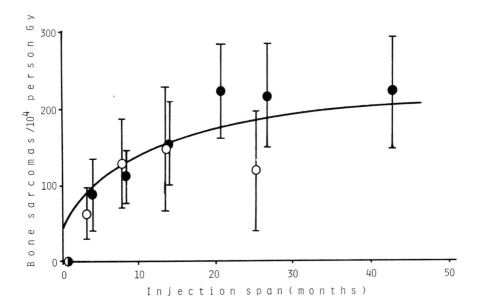

**Figure 9.9. Dose response data for bone sarcoma risk versus protracted doses
of radium-224**
Protraction in children (●) and adults (○)
Source: H. Speiss and C. W. Mays, 1973, in *Radionuclide Carcinogenesis, U.S. Atomic
Energy Agency Symposium Series,* **29,** 442; courtesy the authors.

Numerous surveys have been conducted into carcinogenesis in man. It is from such surveys that we obtain both qualitative information on and quantitative estimates of human radiogenic cancer risks. Before giving these risk estimates it will be instructive to give details of two such surveys.

Ankylosing spondylitics

In the 1950s there was concern in the UK that the radiotherapy given to patients with ankylosing spondylitis† might have very serious long-term effects. Consequently, several epidemiological surveys were carried out and to date 14 111 spondylitics treated with X-rays between 1935 and 1954 have been followed and compared with a control group of 1021 spondylitics not given X-ray treatment. Significantly more cancer deaths were recorded in all sites that were heavily irradiated, for example, the oesophagus, stomach, lungs, spinal cord and nerves. No real excess in cancer was recorded in sites that received minimal doses from the scatter of the X-ray beam that was primarily directed at the spinal region.

The excess mortality from leukaemia occurred between three and eight years after X-ray treatment, with no excess recorded 20 years or more after treatment. In contrast, excess mortality from other cancers was maximal between nine and 20 years with a few cancers appearing at latent periods greater than 20 years. The relationship between leukaemia incidence and dose was linear for mean spinal bone marrow doses between less than 1 Gy and 6 Gy. The data indicate that a total of 10 fatal leukaemias would be expected per million persons per cGy over a 15 year period, with an upper limit of 13 cases per million per cGy.

Childhood cancer from irradiation *in utero*

A survey of cancer in children in the UK was started in 1954 to discover why there had been a progressive increase in leukaemia deaths in young children in the inter-war years. The study covering the period 1953–1967 involved a retrospective survey by medical officers interviewing mothers of 8513 children who had died, mostly under the age of 10 years, from a malignant disease. An equal number of mothers whose children had not died were also interviewed. All mothers were asked about their radiological history during pregnancy and their answers checked with hospital records. The outcome of this UK survey and another in the USA is that it is now firmly established that obstetric diagnostic radiography

---

†Arthritis of the spine leading to complete stiffness of the back.

causes an increased risk of malignant cancer in childhood. The overall estimate of this risk is that there will be some 200–250 cases of fatal malignant disease in children under 10 years old as a result of the antenatal exposure of one million human embryos and foetuses to 10 mGy. Of the 20–25 malignant cancers per year half will be leukaemia and half will be solid tumours. Furthermore, the excess fatal cancer risk increases linearly with the number of X-ray exposures, i.e. from doses between 2 mGy and 200 mGy. The relative overall risk of such radiography was about 1·47, i.e. 47 per cent more fatal cancers will occur in the children of mothers who are radiographed than those that are not. In the UK study the excess risk was found to be five times greater for radiation given in the first three months of pregnancy than for the second and third trimesters. So, a dose as low as 10 mGy given in the first few weeks of pregnancy might cause cancer in 1 in 1000 children. The natural prevalence of fatal cancer in children under 10 years old is 1 child in 2000. These findings were initially greeted with much scepticism but they are now widely accepted as part of the increasing body of evidence that low doses can cause cancer.

Epidemiological data such as that just mentioned are too imprecise to determine the exact shape of the dose response relationship for human radiogenic cancer. Also it is usually necessary to estimate the cancer risk at low doses (<0.01 Gy) by extrapolation of observations made at high doses (>1 Gy). Such extrapolation over one or two orders of magnitude adds to the uncertainty of the risk estimates.

Nevertheless, quantitative risk assessments have to be made and have been made by a number of national and international committees including the International Commission on Radiological Protection (ICRP), the United Nations Scientific Committee on the Effects of Atomic Radiation (UNSCEAR) and by the US Academy of Sciences in their Biological Effects of Ionizing Radiations (BEIR) Reports. Table 9.4 gives the risk rates for fatal cancer that are used by the ICRP for low LET radiation. These rates are a measure of the probability of fatal cancer induction per unit of dose. For low LET radiation the dose units,

Table 9.4. ICRP risk factors for the induction of cancer

Cancer	*Death rates per sievert†*
Leukaemia	$2·0 \times 10^{-3}$
Breast	$2·5 \times 10^{-3}$
Lung	$2·0 \times 10^{-3}$
Bone	$0·5 \times 10^{-3}$
Thyroid	$0·5 \times 10^{-3}$
All other tissues	$5·0 \times 10^{-3}$
Total risk	$1·25 \times 10^{-2}$

† See also table 11.6

sieverts, used in the table are equivalent to a dose in gray. These factors are intended to apply irrespective of age and sex and since men are not susceptible to breast cancer the factor for females is twice that given in the table. The total mortality risk for cancer according to ICRP is $1·25×10^{-2}$ (1 in 80). This means that for every 80 people exposed to 1 Gy of low LET X- or γ rays, one would subsequently die from radiation-induced cancer. The latent period before the appearance of the cancer will vary with the type of cancer and may be as short as 2–3 years for leukaemia or might be more than 20 years for skin and lung cancer. The total cancer risk incidence (fatal and non-fatal cancers) is probably 2–3 times the fatal risk rate. Despite the apparent precision of these internationally agreed risk factors they are only estimates, and should be seen in conjunction with other risk figures. Table 9.5 shows the wide range of cancer risk figures given by some other committees. The ICRP figures are at the lower end of the ranges. The final estimate, 'without the bomb data' is one that excludes the Japanese data. It is reassuring that if one excludes the bomb data, because of the current uncertainty over the dosimetry (see above), the ICRP risk figures are not increased by more than a factor of two. The significance of cancer risk estimates are discussed in relation to radiation protection in Chapter 11.

Table 9.5. Average cancer death risk per 10^4 persons per sievert of low dose rate, low LET radiation

Committee	Range of estimate
BEIR II 1972	115–620
UNSCEAR 1977	75–175
ICRP 1977	125 (no range given)
BEIR III 1980	158–501
UNSCEAR (without the bomb data)	100–440

Despite intensive research over many decades there remain important gaps in both our theoretical and practical understanding of radiation carcinogenesis. Our ignorance is highlighted by the fact that we do not have answers to the following important questions. Is there a threshold dose for cancer in some tissues and not others? How do dose rate and radiation quality affect the process of radiation carcinogenesis? How applicable to man are the dose rate and other studies in animals? To what extent do non-carcinogenic secondary factors play a role in radiation carcinogenesis? While such questions go unanswered there will be both doubt surrounding the cancer risk estimates and an urgent need for more research.

9.7. Radiation biology and radiotherapy

It is a curious fact that even though radiation causes cancer it is also used to eradicate tumours in man, although it is only one of several methods. Cancer therapists also use surgery, chemotherapy (drugs), immunotherapy (enhancing the body's own defence mechanisms) and hyperthermia (heat treatment). These methods are often used in various combinations.

The radiotherapy of cancer has developed empirically, i.e. practice and experience rather than the application of scientific principles have been the guides to treatment. This is quite simply because, even today, no set of scientific rules exists for the preferential destruction of tumour tissue relative to normal tissue. It is one of the practical aims of radiation biology to supply some scientific facts which the therapists may use so that they may obtain optimal results with cancer patients.

In this section we shall draw together a few of the facts that have been described earlier and indicate how they might apply to the radiotherapy of cancer.

From the knowledge you will have gained from this book it may seem easy to cure a tumour. We know that radiation kills cells and that it stops cell division. The inhibition of the proliferative activity of a tumour is equivalent to its cure. Tumours are lethal because they go on growing and eventually may obstruct, for instance, the urinary and intestinal tracts, vital blood vessels, nerves or other tissues. Besides such mechanical pressure and obstruction, infection and haemorrhage are frequent causes of death in cancer patients. A tumour that is localized in the body may be harmless even though it is proliferating. Such benign tumours are in contrast to malignant tumours, which are harmful because not only do they proliferate but they spread or metastasize throughout the body. It is often this systemic spread of cancer that makes it difficult to eradicate.

One might imagine that to cure a cancerous lump composed of one hundred million cells it is simply necessary to give a dose which according to mammalian cell survival curves will kill at least this number. Although this may be theoretically true it is also unfortunately true that there is more to the cure of tumours than cell survival curves—they are of limited clinical value.

In order to ensure that the whole tumour is in the field of the radiation beam, the normal tissues in front of and behind the tumour will receive a certain amount of the dose. By ingenious physical methods it is possible to get most of the dose deposited in the tumour itself, but the normal tissues in the 'treated volume' are rarely left undamaged. Since radiation injures normal and tumour cells with equal ease, the problem that faces the therapist is how to minimize the damage to the normal tissues while inflicting maximum damage on the tumour. It is the unacceptable

Biological Effects of Radiation

damage to the normal tissues that prevents the therapist from merely applying survival curve data and killing the tumour outright with a massive dose.

The most critical normal tissues include the intestinal tract, the lens of the eye, kidneys, lungs, testes and ovaries.

Many years ago radiotherapists learnt by experience that if small radiation doses were given at daily intervals, less damage was done to the normal tissues; they seemed better able to recover between the daily fractions than the tumour.

The therapeutic advantages of fractionation seem to depend on a number of complex factors that we shall now touch upon.

A standard course of radiotherapy is given as daily fractions of 2 Gy, 5 days per week, i.e. 10 Gy per week, for 4-6 weeks. Such treatment is

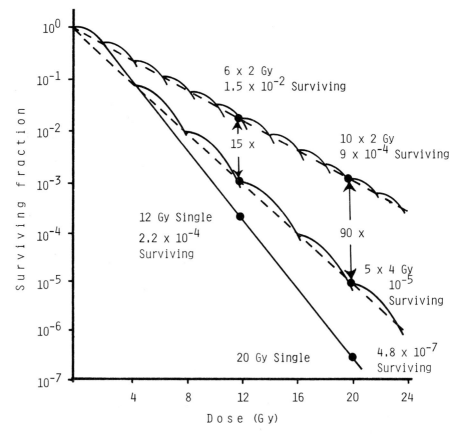

Figure 9.10. Diagram of cell survival from single doses, fractions of 4 Gy each dose and fractions of 2 Gy each dose
Source: W. Duncan and A.H.W. Nias, 1977, *Clinical Radiobiology*, Churchill Livingstone, courtesy the authors and publisher.

most likely to be from supervoltage machines like ^{60}Co γ ray units or linear accelerators, both of which deposit their maximum dose just below the skin and so avoid too much damage to the skin. The tolerance dose of the skin for fractionated treatment is about 60–70 Gy. The term tolerance dose is a clinical judgement and usually implies the dose that will produce no more than about 5 per cent of serious consequences. Clinical tolerance doses range from about 5 Gy for the ovary and the bone marrow to as high as 150–200 Gy for the bile duct, with most tissues having intermediate values of about 30–80 Gy.

The effect of fractionating the dose on cell killing was discussed in Chapter 4. We saw that splitting the dose into small fractions divided by time intervals allowed cell recovery and was therefore an inefficient way of killing cells relative to single large doses. Figure 9.10 illustrates this reduced effectiveness of multiple doses compared with single doses. The figure shows two fractionation regimes both assuming the complete restoration of the shoulder of the survival curve between doses. It is obvious that a series of smaller doses is even less effective than a series of larger fractions. The figure illustrates one of the crucial time dose factors therapists must consider.

The so-called five 'R's of radiobiology are often invoked to explain why radiotherapy works. These 'R's are *R*adiosensitivity, *R*ecovery, *R*eoxygenation, *R*epopulation and *R*edistribution of cells in the cell cycle. Coupled with these factors there are such physical factors as dose rate, quality (LET) of the radiation, dose distribution and dose–time factors (the number, size and interval between fractions and the overall treatment time).

There are really no significant differences between the radiosensitivities for the killing of normal and tumour tissues. The minor differences that are sometimes apparent are insufficient to explain the efficacy of radio therapy. With the important exception of patients with Hodgkin's lymphoma, where the cancerous lymphocytes are so much more radiosensitive than most normal tissues that over 80 per cent of patients will survive for more than 5 years after combined radiotherapy and chemotherapy. Untreated, the disease is almost inevitably fatal within two or three years of diagnosis.

Similarly there are few differences between the repair abilities of normal and tumour tissues as measured by the quasi-threshold (D_q) in experiments *in vitro*. However, it is possible that since many tumours are hypoxic *in vivo* or even temporarily anoxic (see below) and since repair mechanisms are oxygen dependent (see Chapter 8) then the accumulated damage in tumour tissues of the multiple doses may be greater than the damage suffered by normal tissues.

Reoxygenation is an important phenomenon that occurs in most tumours but not normal tissues. It is known that as tumours enlarge they outgrow their blood supply. The diffusion distance of oxygen in tissues is about 150 µm, so cells more than 150 µm from a capillary or other blood vessel will become anoxic as they metabolize any available oxygen. Minute areas of dead and dying cells—'hypoxic foci'—are probably present in most tumours more than 2 mm in diameter. The proportion of hypoxic or even anoxic cells may be as much as 20 per cent or more. Following radiation the oxygenated (radiosensitive) cells will be preferentially killed and the proportion of hypoxic cells may even reach 100 per cent. However, as the radiation-killed cells are cleared away by phagocytosis the tumour shrinks, the tumour blood vessels, i.e. the oxygen supply, will be relatively unaffected by the radiation, and so the remaining tumour cells will suddenly become well oxygenated and again become radiosensitive to subsequent doses. Unfortunately, little is known of the time sequence or magnitude of the fluctuating pattern of reoxygenation in human tumours.

Regrowth or repopulation of tissues can occur between depopulating doses of a fractionation regime. If the amount of this cell proliferation were greater in normal tissues than in tumour tissues then the latter would gradually regress and the normal tissue might maintain its normal cellularity. There are experimental grounds for believing such differential

Centimetres

0   5   10

Theoretical dose for eradication of 100% well oxygenated tumour

24   27   31   35   38   41   45

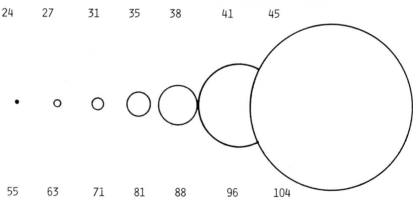

55   63   71   81   88   96   104

Theoretical dose for eradication of 100% anoxic tumour

Figure 9.11.   The influence of tumour size on the theoretical dose (in grays) of radiation required for tumour eradication

Source: H. B. Hewitt, 1962, *Scientific Basis of Medicine: Annual Reviews,* University of London Athlone Press; courtesy the authors and publisher.

regrowth may occur but it is probably less important than the recovery factor.

Finally, there is the possibility of redistribution of the cells in the cell cycle. In Chapter 8 we noted that the radiosensitivity of cells varies through the cell cycle. It is suggested that fractionated doses will preferentially kill cells in the most sensitive phases. A differential reassortment of tumour cells relative to normal cells might have the effect of prefentially depopulating the tumour. Many workers find this a doubtful proposition.

The presence of small numbers of hypoxic (radioresistant) cells may make all the difference between the success and failure to cure a tumour. This is illustrated in figure 9.11, which shows the radiation dose needed to kill all the cells in the volumes indicated. Much lower doses are required for fully oxygenated (radiosensitive) tumours, and it is easy to see the dramatic difference that hypoxia or anoxia make.

In Chapter 8 it was mentioned that the blood supply to most, although not all, normal tissues carries enough oxygen to ensure that they are maximally radiosensitive, whereas some parts of some tumours are hypoxic and therefore radioresistant. This is shown in figure 9.12. This differential radiosensitivity means that a greater proportion of normal cells than tumour cells will be killed by a given dose. This is most undesirable. A number of methods have been tried to deal with hypoxia in tumours. These include hyperbaric oxygen, tourniquet hypoxia, high LET radiation and the use of drugs called hypoxic cell sensitizers.

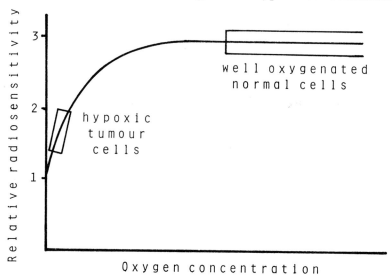

Figure 9.12. Illustration of the relative radiosensitivity of well oxygenated normal tissues and the hypoxic areas in tumour tissues

If a patient is put in a high-pressure oxygen (HPO) chamber, the oxygen concentration of all the tissues rises. The tumour will therefore become more radiosensitive, and since the normal tissues are already maximally sensitive, the differential killing may be decreased. In figure 9.12 what HPO does is to cause the anoxic cells to move in effect up the curve towards the sensitivity of the normal tissues.

The results of recent hyperbaric oxygen trials with radiotherapy are promising, especially in the local control of carcinoma of the larynx and the cervix of the uterus. However, the five-year survival rates were not encouraging.

Alternatively, where it is possible, a tourniquet can be applied so that the oxygen content of all the tissues falls—the reverse of the HPO effect. This will reduce the preferential killing of the normal cells since it will cause their sensitivity to move down the curve in figure 9.12 towards the already hypoxic tumour cells.

The third method to reduce the preferential and undesirable killing of normal cells is the use of high LET radiation.

Neutrons, negative π mesons (pions) and light atomic nuclei are examples of densely ionizing nuclear particles that are beginning to play a role in cancer treatment. Such radiations have three main advantages over low LET radiations in radiotherapy. There is less repair of their radiation damage; less variation in radiosensitivity in the phases of the cell cycle; and most importantly, the OER of high LET radiations is between 1 and 2 compared with 2·5–3·0 for low LET X- and γ rays. These phenomena are all covered in Chapter 8.

Fast neutron beams obtained from cyclotrons are now in use in a dozen countries.

The initial results of clinical trials are encouraging for the local control of some carcinomas of the head and neck. The trials are still at an early research and development stage and years of long-term clinical work will be required before one could state that neutrons are the preferred cancer treatment.

Beams of negative pions (sub-atomic particles 276 times heavier than electrons) are produced in extremely high energy accelerators by bombarding copper, tungsten and other 'targets' with 500–750 MeV protons. The pions produced deposit their energy mainly in the Bragg ionization peak (see figure 1.2), the entrance dose to the skin and normal tissues being relatively low compared with the ionizations at the peak in deeply sited tumours. The OER of pions is much less than that of low LET X- or γ rays. To date less than 100 cancer patients have been treated with pions, and there are clinical facilities only in Canada, Switzerland and the USA.

Accelerated beams of light atomic nuclei, for example helium, carbon, neon and argon, have similar physical properties to pions—OERs of slightly greater than 1·0, high RBE values and depth–dose distributions in tissue that allow Bragg peak irradiation of deep-seated tumours. All these high LET particulate radiation beams are expensive to produce, difficult to shield and collimate, and only fast neutrons have reached the stage of being tested in proper clinical trials.

Finally, and perhaps most promising, a group of drugs has been developed that may overcome the radioresistance of hypoxic tumour cells. The chemicals, called 'hypoxic cell sensitizers', may eventually lead to a great improvement in the ability of radiation to cure local tumours (see Chapter 8 for details).

Numerous clinical trials are underway which, if they confirm the research results, will produce great improvements in local tumour control rates. Currently, neurotoxicity limits the total drug dose to as little as 10 per cent of that needed to obtain maximum radiosensitization of hypoxic cells. However, there is promising experimental work with newer drugs that are less able to cross the blood–brain and blood–nerve barriers. It seems quite possible that a future generation of such sensitizers may obviate the need for expensive high LET radiation.

There has recently been a great deal of research interest in the combination of hyperthermia and radiation to cure cancer. The combination of heat (about 43 °C) and radiation on animal tumours and cells in culture has produced some interesting results. It has been found that (a) cancer cells are more sensitive to heat than normal cells; (b) hypoxic cells are more sensitive to heat than oxygenated cells; (c) a combination of heating and radiation kills more tumour cells than either of the treatments alone; and (d) cells in DNA synthesis which are normally radioresistant (see Chapter 4) are the most sensitive to heat treatment. Why a combination of heating and radiation is synergistic in killing tumour cells is not fully understood but the mechanism probably involves a reduction in the capacity of the heated cells to restore DNA damage. Current research suggests that the best clinical results may be obtained by a period of local pre-heating of the tumour before the radiation followed by a slightly longer period (1–2 hours) of post-heating.

Hyperthermia, like hypoxic cell sensitizers, is at the research stage and there are reports in the literature that contradict many of the claims made for hyperthermia. There are technical difficulties in accurately and locally raising and maintaining the temperature of a tumour. Further, there is a problem of thermal tolerance—it seems that tissues become resistant to repeated heating.

A final aspect in which radiobiology may one day be able to rationalize the radiotherapy of cancer is to use the differential sensitivity of the phases of the cell cycle (see Chapter 4). However, at present there is little evidence that the fluctuations in tumour cells differ from those in normal cells, nor is there any way as yet of preferentially marshalling the cells of the tumour into a radiosensitive phase.

9.8. Summary

Cancer induction is the most intensively studied and the most significant late effect of radiation. The most favoured hypothesis is that radiation causes cancer by inducing somatic mutations. It is capable of inducing tumours in almost all the tissues of the body although the tissues vary greatly in their susceptibility. Great empirical and theoretical efforts have been made to understand the fine detail of the relationship between cancer incidence and dose, but despite this, the simplest relationship, i.e. linearity, is still held to be consistent with most of the experimental and human data, at least at low doses. At higher doses, many experimental animal cancer incidences plateau or even show peak values that suggest a competitive effect between cancer induction and cell killing.

There have been numerous epidemiological studies of human radiogenic cancer and from these internationally agreed risk estimates have been obtained. Nevertheless, until we know more about cancer in general and radiation cancer in particular such risk estimates will remain tentative and subject to change.

Radiation can also be used to cure cancer. Radiation cell killing and the inhibition of cell division were discussed at length in Chapters 3 and 4. Unfortunately, the killing of cancer cells is not quite as simple as merely applying data from survival curves. The use of single large doses of radiation, designed to kill all tumour cells, causes unacceptable damage to the normal tissues surrounding the tumour. Radiotherapists have learnt by experience that the use of a series of smaller daily fractions is the only way to eradicate tumours without excessive damage to the normal tissues. Such dose regimes, besides allowing the repair of normal tissues, also allow the repair of tumour tissues and so are an inefficient way of killing the tumour; but at present it is apparently the only way of successfully treating tumours with radiation. The success of therapy seems to rest upon a number of factors that produce the complex of differential responses between the cell proliferation of normal and tumour cells. These factors include radiosensitivity, repair, repopulation, reoxygenation of cells, and redistribution of cells within the cell cycle.

Tumours often contain a number of hypoxic, radioresistant, cells that make them more difficult to cure. Normal tissues, on the other hand, are usually fully oxygenated, i.e. radiosensitive. This difference means that a greater number of normal cells than tumour cells will be killed by the radiation. The methods to reduce this undesirable effect include the use of high-pressure oxygen, tourniquet hypoxia, high LET radiation and drugs called hypoxic cell sensitizers.

Chapter 10
Late effects of radiation

10.1. Late effects in general

We have mentioned that radiation causes not only early or acute effects in tissues but also causes late or delayed effects. The latter include genetic, teratogenic and carcinogenic effects and these have been discussed in Chapter 7, Chapter 6 section 9 and Chapter 9, respectively. We saw in Chapter 6 that the acute effects of radiation generally occur within days or weeks and in rapidly dividing cell populations such as the skin, the bone marrow, and the gastrointestinal tract. The damage has been described as a 'disturbance of cell kinetics' and it is due to the mitotic death of cells in such rapidly proliferating populations, providing the doses are not too large the tissues are rapidly repopulated and repaired and little permanent damage is left.

In contrast, slowly proliferating or non-proliferating tissues, for example lung, liver, kidney, heart, connective tissue, nervous tissue and bone, exhibit a diversity of late radiation damage. There are two schools of thought concerning the cause (pathogenesis) of these late effects. One hypothesis suggests that late damage, i.e. months, years, or even decades after the radiation, is primarily due to vascular damage. It is thought that damage to small blood vessels eventually leads to a degeneration of the cells of a tissue and to a generalized late fibrosis of its associated connective tissues. The other school of thought suggests that the wide diversity of late effects seen in different tissues, the differences in the time of their appearance and of their severity and rate of development, are best explained in terms of cell kinetics. This hypothesis suggests that just as acute radiation effects occur early in rapidly dividing tissues, so late effects occur late in slowly- or non-proliferating tissues. The effects are 'late' because the expression of the damage at mitosis has to await the slow appearance of such divisions. Of course, the truth may be that late effects are due to a combination of effects in the connective tissues, the vascular tissues and the parenchymal tissues.

10.2. Specific late effects

It is not possible to describe in detail the wide variety of late effects in all of the tissues and we shall content ourselves with one or two important instances.

The eye

The eye is a highly complex structure composed of a variety of tissues ranging from the relatively radiosensitive parts such as the lens, through moderately responsive ones such as the cornea and conjunctiva, to radioresistant ones such as the optic nerve and the retina.

Radiation damage to the cellular fibres that comprise the lens of the eye can cause opacities. If these are large they can interfere with vision and are termed 'cataracts'. The dose response curve for the induction of lens opacities is sigmoid for both low and high LET radiation. In man, the threshold dose to produce opacities in a few per cent of those exposed is about 2 Gy of acute γ or X-rays. Doses of 6–7 Gy would cause cataracts in most of those exposed and would be likely to seriously impair vision. For high LET radiation the RBE for cataracts is probably between 3 and 10. The time interval between irradiation and the appearance of the opacities is highly variable and may be as short as 6 months or as long as 30 years, with a median time of 2–3 years.

Since the lens contains no blood vessels this late effect is due solely to damage to the cells that comprise the lens.

The skin

Both the acute and delayed effects in the skin of man are well documented from clinical practice. Acute 'radiation dermatitis' was described in Chapter 6. The late effects in the skin include the contraction and atrophy of the irradiated area, its loss of elasticity, the non-functioning of sweat glands and hair follicles, dermal fibrosis and delayed healing of the skin following any trauma. This complex of late injuries is likely to be due to a combination of cell depletion, vascular damage and fibrocyte dysfunction (causing fibrosis). Local doses of about 10–20 Gy are required for significant late effects in the skin. The severity of the late effects in the skin depend on a number of factors including the volume of tissue irradiated, the radiation quality and the time over which the dose is administered. There seems to be little correlation between the severity of the early effects and the severity of the late effects.

The connective tissue

The effects just mentioned in the skin are a good example of the most common late effect seen in radiotherapy patients, 'late fibrosis'. The dermal fibrosis in skin involves the replacement of normal dense connective (elastic fibres, collagen bundles, smooth muscle fibres, etc.) with a relatively uniform mass of acellular collagen. This type of late fibrosis develops in most heavily irradiated tissues, for example lung, kidney, liver, and so on. As mentioned above, fibrosis may be due to vascular injury which would cause areas of ischaemia. Alternatively, the slow loss of the fibroblasts that are responsible for the continual secretion and absorption of collagen and other fibres would lead to the quasi-crystallization or 'premature ageing' of the existing collagen.

The lung

The lung is a complex organ in which more than 40 different cell types have been described. Besides the epithelial cells of the conducting airways and of the distal respiratory alveoli, there are numerous other cells—muscle cells, nerve cells, blood vessel and lymphatic elements, and so on.

Radiation pneumonitis was mentioned in Chapter 6 and this occurs within 4–6 months of fractionated doses of 40–50 Gy. The main late effects in the lung include severe fibrosis of the alveolar walls and a progressive sclerosis and loss of the fine vasculature of the respiratory tissues. Such changes gradually cause a reduction in the physiological 'functional reserve capacity' of the lung and a proneness to pneumonias and other infections. Such clinical problems may not become apparent for many years after radiotherapy.

Many of the above effects, together with late damage to the spinal cord, peripheral nerves and kidney, have significance for radiation tolerance in radiotherapy (see Chapter 9 section 7). Compared with our knowledge of the acute effects of radiation our understanding of late effects is still rudimentary. Much remains to be discovered of the basic pathobiology of both cancerous and non-cancerous late radiation damage.

10.3. Radiation lifeshortening

The shortening of life by radiation is a well-documented phenomenon. However, before one can begin to study it one must have a detailed knowledge of the pattern of mortality in a population that has received no radiation (controls). Figure 10.1 is a typical curve for the percentage of

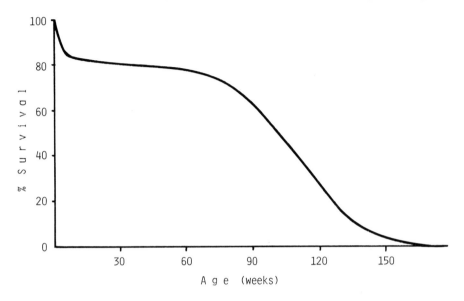

Figure 10.1.   Survival curve from birth of unirradiated mice
Source: P. J. Lindop, 1965, *Scientific Basis of Medicine: Annual Reviews,* p. 91; courtesy the author and Athlone Press.

control mice surviving to a given age. It can be seen immediately that there is a small but sharp fall around the first few weeks of life, after which the probability of survival remains almost constant until about 90 weeks of age. From then onwards the mice begin to die in ever-increasing numbers, i.e. the probability of dying rapidly increases after a certain age is reached. Eventually only a few very old mice are left. This type of 'square' survival curve is typical for many unstressed populations of animals, including man in the developed countries. It will be obvious that many hundreds of mice are required for an accurate determination of a population survival curve.

Figure 10.2 shows the effect of giving two different doses of radiation to two very large populations of mice. As the dose gets larger the mice tend to die at an earlier age. Radiation is causing a shortening of the average lifespan of the population. This statement can be seen graphically in figure 10.3 which shows lifeshortening as a function of dose for mice of

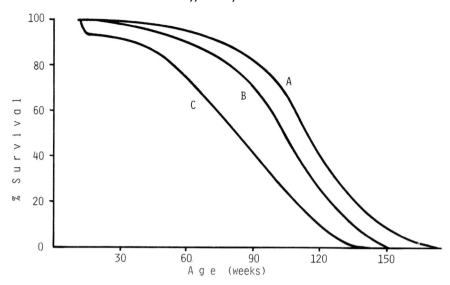

Figure 10.2. **Survival curves of populations of 4 week old mice given no irradiation (curve A); 1.98 Gy (curve B); or 4·57 Gy (curve C) of 14 MeV electrons**
Souce: P. J. Lindop, 1965, *Scientific Basis of Medicine: Annual Reviews*, p. 91; courtesy the author and Athlone Press.

both sexes. The relationship appears to be linear down to a dose as low as 0·5 Gy.

In figure 10.3 a line drawn through the points goes through the origin of the graph. This suggests, but does not prove, that however small the dose of radiation, it will have some lifeshortening effect. This raises the problem of whether a threshold dose exists for lifeshortening.

A recent lifespan study using about 27 000 mice and whole-body γ ray doses at 0·45 Gy min⁻¹, reported that a linear fit was not possible at doses between 0·1 and 0·5 Gy. At these low doses the best fit was a dose square function. Since this is the largest experiment to date, and unlikely to be repeated, some weight must be given to its findings. However, since the amount of lifeshortening reported in the experiment was so much greater than that of other studies, some reservations must remain. So, at very low doses it is not possible to decide the exact shape of the relationships between lifeshortening and dose, and until this information becomes available it is best to err on the side of safety and assume that any dose of radiation is capable of shortening life. The problems surrounding low dose risks were discussed at length in Chapter 9 for cancer induction.

The slope of the line drawn in figure 10.3 shows that, in 30-week-old mice, for every 1 Gy received their life-expectancy is reduced by about 5 weeks. What amount of lifeshortening do old mice show? As mice get older the number of weeks by which their lives are shortened following a

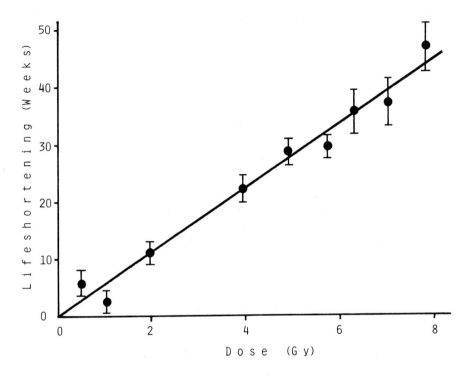

Figure 10.3. Lifeshortening as a function of dose for mice of both sexes
Source: P. J. Lindop, 1965, *Scientific Basis of Medicine: Annual Reviews,* p. 91; courtesy the author and Athlone Press.

dose of radiation decreases. However, if the amount of lifeshortening in older mice is expressed in terms of the amount of lifespan remaining to them, a constant value is obtained. This constant value suggests that a total body exposure of 1 Gy given at any age will reduce the remaining lifespan by 5 per cent. At very advanced ages no lifeshortening is observed, which suggests that there is an induction period of some 5–6 months before the delayed effect of lifeshortening shows itself. The value of about 5 per cent lifeshortening per 1 Gy of low LET radiation has been found to apply to numerous strains of mice in a dozen separate experiments.

So far, only the effects of acute doses of X- and γ rays on lifespan have been discussed. There is much less information for mice exposed at very low dose rates. Several studies have been reported in which mice were exposed throughout their lives to daily doses of γ rays ranging from 0·003 to 0·56 Gy. The main finding of such protracted radiation exposure was that mean survival times after the initiation of the exposures declined exponentially with the daily dose d (in gray) according to the equation

$$\text{MAS (treated)} = \text{MAS (control)} \ e^{-0\cdot04\,d}$$

where MAS stands for 'mean after survival'. At the lowest daily doses no significant lifeshortening was detectable.

A comparison between the effectiveness of acute versus protracted low LET exposure suggests that the protracted doses are some five to ten times less effective in producing lifeshortening in the mouse. This comparison assumes linearity of lifeshortening with doses for both acute and protracted exposure. In contrast, low dose rate neutrons are almost as effective as single acute neutron doses.

Most of the experimental lifespan studies have been on mice although the data from dogs, rats and guinea pigs are in qualitative agreement with the conclusions reached for mice. Besides the experimental data there are several pieces of evidence that point to a radiation lifeshortening effect in man. A survey in 1956 of over 80 000 obituaries of US doctors who died between 1930 and 1954 showed that the mean age at death of the radiologists was 60·5 years, in comparison with a mean age at death of 65·7 years for doctors not exposed to daily radiation and a mean age at death of 63·7 years for doctors with an intermediate degree of radiation exposure. This study was criticized on statistical grounds, but its conclusions were later ratified in a much more detailed survey of American medical personnel. This study was in 1965 and showed that the median age at death for radiologists was 5 years less than for doctors not occupationally exposed.

A study of the mortality of UK radiologists failed to reveal any radiation lifeshortening. This discrepancy between the USA and UK data is held to be due to the fact that the latter adopted rigorous standards of radiation protection as long ago as 1922, some 20 years before the USA. It is believed that the modern standards of protection have now reduced occupational exposure to very low levels that are not a significant hazard.

Further indication of lifeshortening has come from the recent reports on the mortality of the Japanese A-bomb victims. The Atomic Bomb Casualty Commission studied the deaths between 1950 and 1960, in a sample of approximately 100 000 survivors of all ages from the Hiroshima and Nagasaki bombs, and compared the deaths with those in a similar number of exposed persons. The Commission found a 15 per cent increase in general mortality of those people who had been within 1200 m of the hypocentres of the atomic bombs. This increase is exclusive of leukaemia deaths. As we saw for radiation carcinogenesis, the data on radiation lifeshortening in man are qualitatively consistent with those for experimental animals. The studies on US radiologists and the atomic bomb victims demonstrate that radiation lifeshortening cannot be attributed solely to cancer induction but may have a non-specific element. The human data are insufficient to construct dose reponse curves for lifeshortening, and the International Commission on Radiological Protection does not feel that the evidence for lifeshortening in man is "sufficient to justify any quantitative estimate of the risk".

10.4. Some theories of lifeshortening

As soon as the fact that radiation causes a reduction of lifespan was established, questions began to be asked as to how radiation might bring about premature death. Was radiation acting like some chemicals and inducing special diseases? Does radiation cause premature ageing and death by causing special types of cancer? Or does radiation advance all the natural causes of death?

Although radiation-induced lifeshortening is an easily measurable effect, it is a most complicated biological process, the precise nature of which is far from understood.

Some of the early experiments strongly suggested that radiation reduced lifespan by causing the premature onset of all the diseases and causes of death that occur in unirradiated animals. So it was suggested that radiation lifeshortening was a result of a non-specific injury to all the tissues of the body. This widespread damage was assumed to cause the premature deterioration of tissues and to increase their susceptibility to the whole spectrum of diseases that characterize natural senescence.

Considerable doubt has been cast upon the interpretation of many of the experiments purporting to show that radiation causes 'non-specific lifeshortening', i.e. advances all the natural causes of death. More recent interpretations suggest radiation lifeshortening is a "biological integrator of radiation carcinogenesis", i.e. premature death is due essentially to the induction of a whole variety of cancers.

Nevertheless, the lifeshortening in two large experiments and in some human data cannot be attributed to cancer alone, since the lifeshortening persists after all the cancer deaths have been discounted. In many experiments the parallelism between natural ageing and the premature ageing after radiation is striking. A closer look at natural ageing might provide some clues as to radiation lifeshortening.

It is unfortunate that there is no universally accepted definition of 'ageing'. Most people would probably accept the rather vague formulation that ageing is the gradual loss of functional capacity and vigour that leads to an increasing susceptibility to disease. This definition tells us nothing of the basic biochemical and biophysical mechanisms of growing old, the reason being that they are not understood in any detail. Despite this ignorance there are several hypotheses of ageing that may be roughly grouped under three headings—the wear and tear hypotheses, the accumulation of toxins, and the somatic mutation theories.

The wear and tear, or stress, hypotheses suggest that all the biochemical, physiological and anatomical injuries an animal receives accumulate throughout its life. Some of this damage will be repaired (wound healing, etc.) and some of it will constitute an irreparable fraction which will ultimately cause senescence and death. One of the most interesting of these wear and tear hypotheses suggests that the free radicals that are produced as intermediaries of metabolism might injure other biological molecules throughout life and lead to ageing; radiation could add to the natural level of injury, thus hastening the onset of ageing and death. If natural ageing is due to free radical damage, then radiation is eminently suited to cause ageing because, as we saw in Chapter 1, it produces vast quantities of free radicals.

The second group of theories suggests that ageing is a result of the accumulation of certain waste substances. The two most important examples are the build-up of chemical bonds between large molecules and the progressive accumulation and denaturation of collagen with age. The idea of the build-up of various cross-linkages (including hydrogen bonds) between the macromolecules of living tissues as the basis of ageing is comparable to the ageing of paper, rubber and natural and synthetic fibres. Although radiation does produce cross-linkages between molecules, these are not detectable at the present time at the dose levels that shorten life. Collagen is an extracellular material that supports, connects and protects cells and tissues, and since it is widespread in the body and changes in its composition or concentration are likely to have far-reaching repercussions. These facts have led some theorists to speculate that the age changes seen in collagen are the basis of natural ageing. If this hypothesis is true it is difficult to explain the remarkable closeness between natural and radiation ageing, because collagen is relatively unaffected by radiation.

The third heading, somatic mutation theories, includes very many hypotheses. The theories all suggest that the molecules that carry vital information in cells are susceptible to gradual deterioration. This will mean that they transmit incorrect messages that will lead to a gradual accumulation of defects in the cells, tissues and organs. This gradual deterioration of function is of course senescence. Since DNA is the main information carrying molecule in the cell, it is not surprising that many people feel that ageing must in some way involve changes in the nucleic acids.

The hypothesis is that the DNA of the somatic cells of the body initially contains a perfect message. Through one process (faulty replication?) or another (environmental stresses?), the message gradually becomes more and more garbled. A faulty DNA message means that protein synthesis, and consequently cell structure and function, will be less than perfect. It is true that the average lifespan of animals is in some way genetically controlled. It is also true that chromosome aberrations in liver cells of mice increase with age, and that an age-related increase in the incidence of aneuploidy in human peripheral blood lymphocytes has been reported, but it is not known whether such abnormalities are the cause of ageing or a reflection of age-related disease states. However, if natural ageing is a result of somatic mutations in the DNA molecules, radiation seems ideally suited to mimic natural ageing since, as we have seen many times in this book, radiation very effectively damages DNA.

The hypotheses above are some of the many speculations that have been made in an attempt to understand the mechanism of ageing. None of them really helps us to understand radiation lifeshortening. Natural ageing is the progressive deterioration of living tissues and radiation causes the premature appearance of such a deterioration, but it is not clear whether the fundamental cause(s) of natural lifeshortening are the same as those of radiation ageing. Neither natural nor radiation ageing can be attributed to a single specific cause and although in some studies radiation lifeshortening may be predominantly due to carcinogenesis, there are other studies where the induction of lesions other than cancers plays a major role. Since we are ignorant of the mechanisms of natural ageing it is hardly surprising that we are ignorant of those of radiation lifeshortening.

10.5. Summary

This chapter has considered the late or delayed effects of radiation, excluding carcinogenesis, teratogenesis and hereditary effects. The late effects in many slowly or non-dividing cell populations, for example the lung, the connective tissues of the skin, bone, kidney, nervous tissue, and so on, are often important limiting factors in clinical radiotherapy. There are two schools of thought on the pathogenesis of such late effects. One suggests that the expression of damage is 'late' because it has to await the slow appearance of post-irradiation cell divisions to reveal the damage. The other school suggests that the major cause of late effects in other slowly proliferating tissues is due primarily to vascular damage.

The chapter also examined the evidence for radiation lifeshortening in animals and man. Acute doses of X- and γ rays produce an approximate 5 per cent reduction in lifespan in most species studied. Protracted low LET radiation is some five to ten times less effective than acute doses at causing lifeshortening. The dose response curve for lifeshortening is probably curvilinear at the lowest doses but some large experiments indicate that linearity may not be ruled out and therefore there may not be a threshold dose for lifeshortening. There is evidence that lifeshortening occurs in man. The major cause of radiation lifeshortening is probably from cancer induction, but there remains strong evidence to suppose that to some extent the lifeshortening reflects the induction of a whole spectrum of non-specific lesions other than cancer.

Chapter 11
Human radiation exposure and the standards for radiation protection

11.1. Introduction

The International Commission on Radiological Protection (ICRP) was founded in 1928 and, broadly speaking, its role is to evaluate the risks that man runs when exposed to radiation and to set limits on the maximum permissible levels of exposure for the general public and for radiation workers. Both the hazards and the recommended permissible dose levels are continually being revised in the light of new research data, and this continual revision has meant that the maximum permissible levels today are much lower than they were in 1928.

Before outlining the recommendations of the ICRP concerning radiation protection it will be useful to have an idea of the levels of radiation to which man is more or less continually exposed.

11.2. Natural sources of radiation

Table 11.1 lists the major components of the natural sources of radiation. The estimated doses in μGy per year are given for the gonads, lung, bone marrow and endosteal cells lining the bone surfaces. These tissues are especially important in the assessments of hereditary damage, leukaemia induction and lung and bone cancer.

The values in the table are taken from the 1977 UNSCEAR Report. They are estimated mean values for the world's population. Cosmic radiation from outer space interacts with the atoms of the upper atmosphere and produces electromagnetic and particulate radiation on Earth. The dose rate on the ground varies with altitude, latitude and longitude and the values given are averages. Terrestrial radiation comes from the naturally occurring radio-isotopes that are present in all the

Table 11.1. Dose rates from natural radiation sources (μGy yr^{-1})†

Source	Gonads	Lung	Bone marrow	Bone-lining cells
Cosmic radiation				
Ionizing component	280	280	280	280
Neutron component	3·5	3·5	3·5	3·5
Terrestrial γ radiation				
(including the air)	320	320	320	320
Internal radiation				
^{40}K $(\beta+\gamma)$	150	170	270	150
^{222}Rn–^{214}Po (α inhalation)	2	300	3	3
^{210}Pb–^{210}Po $(\alpha+\beta)$	6	3	9	34
^{87}Rb (β)	8	4	4	9
^{14}C (β)	5	6	22	20
^{3}H (β)	–	0·01	0·01	0·01
Total	**~780**	**~1100**	**~920**	**~860**

Source: The Report of the United Nations Scientific Committee on the Effects of Atomic Radiation (UNSCEAR) 1977. Official Records of the General Assembly 32nd Session Supplement No. 40 (A/32/40).
†Minor sources of internal radiation are not included.

rocks and soils of the Earth and therefore of the world's building materials. Most of the radionuclides that were present at the time of the formation of the Earth have of course decayed leaving only those with extremely long half-lives $(t_{\frac{1}{2}})$. The commonest of these isotopes in the Earth's crust are uranium-238 | $(t_{\frac{1}{2}}=4\cdot5\times10^{9}$ years), thorium-232 $(t_{\frac{1}{2}}=1\cdot4\times10^{10}$ years) and potassium-40 $(t_{\frac{1}{2}}=1\cdot28\times10^{9}$ years). The γ rays emitted by such radionuclides irradiate the whole body more or less uniformly. Internal radiation is produced by the ingestion and inhalation of naturally occurring radionuclides, of which potassium-40, rubidium-87, lead-210, radium-226 and radium-228 are important. However, the most significant source of internal radiation is from the inhalation of the radioactive gas radon-222 and its radioactive decay products. Radon $(t_{\frac{1}{2}}=3\cdot8$ days) enters the atmosphere from the Earth's crust or the floors and walls of buildings and decays rapidly by α emission. The 'radioactive daughters' so produced attach themselves to dust particles and are readily inhaled and irradiate the lung. A final source of internal terrestrial radiation comes from radio-isotopes of hydrogen-3 (tritium), carbon-14 and beryllium-7 that are all formed in the air by neutron interaction from cosmic rays.

It should be noted that table 11.1 lists only the major sources of internal radiation and the overall rounded totals of approximately 1000 μGy to each of the organs are for people living in areas of average radiation background. There are areas in Brazil, France, India and Iran, for example, where the radioactive content of sands, rocks and soils is several times higher than average.

11.3. Artificial sources of radiation

The natural sources of radiation affect the whole of the world's population all of the time, whereas the artificial sources, for example medical and occupational exposure, only affect a small fraction of the population at any one time. Table 11.2 is a list of the typical bone marrow doses from some medical X-ray diagnostic examinations. Since many such examinations involve some irradiation of the gonads (see table 11.3) there is considerable interest in the hereditary (genetic) defects that might be caused.

Table 11.2. Typical bone marrow doses during diagnostic X-ray examinations

	Median dose (μGy)	
Examination	*Male*	*Female*
Barium meal	5 100	8 000
Mass survey chest	610	1 000
Lumbar spine	2 700	2 700
Abdomen	1 200	1 300
Descending urography	5 800	4 500
Retrograde urography	4 400	3 300
Dental	18	18
Abdomen obstetric	–	2 100[a]
Pelvimetry	–	2 800[b]

Source: United Kingdom Committee on Radiological Hazards to Patients 1966, Final Report of the Committee (H.M.S.O.).
[a]Dose to foetus 5 000 μGy.
[b]Dose to foetus 11 000 μGy.

The data in both tables 11.2 and 11.3 are median values taken from surveys in a number of countries and therefore conceal wide variations. For example, the median values in table 11.3 for gonadal doses from radiological examinations of the lumbar spine are 2100 μGy for males (range of mean values 260–22 700 μGy) and 4100 μGy for females (with a range of 2300–11 900 μGy). A survey in the UK in 1977 also noted the enormous variability in gonadal doses between different hospitals. The survey concluded that such variability indicated that some patients are receiving unnecessarily high doses and stressed the need for a more consistent use of gonad shielding especially in young adults.

The 'genetically significant dose' (GSD) for a population as a whole is that dose which, if received by all members, would produce the same genetic effect as the actual doses received by the gonads of the particular individuals that undergo radiological examination. To estimate this genetic dose for the population it is necessary to make allowance for the fact that a proportion of the people exposed will not produce offspring and will not therefore pass on any radiation-induced genetic damage.

Table 11.3. **Typical gonad doses to adults from diagnostic X-ray examinations**

	Median dose (μGy)	
Examination	*Male*	*Female*
Barium meal	300	3 400
Mass chest survey	4	30
Lumbar spine	2 100	4 100
Abdomen	2 500	4 100
Descending urography	4 300	5 900
Retrograde urography	5 800	5 200
Dental	6	0·6
Abdomen obstetric	–	3 000
Pelvimetry	–	6 200

Source: UNSCEAR Report, 1972, Official Records of General Assembly 27th Session Supplement No. 25 (A/8725).

These weighting procedures for the individual's child-bearing expectancy are part of the GSD calculation. The GSD in Great Britain in 1977 was estimated to be 0·114 mGy compared with about 0·2 mGy in the USA and Japan, about 0·3 mGy in Italy, Romania and Netherlands, 0·4 mGy in West Germany and Switzerland and 0·45 mGy in Sweden. In the case of occupational exposure GSD is of lesser importance because of the age distribution of the population.

As with medical exposure, occupational exposure to radiation affects only a small number of individuals, such as medical, dental, atomic energy, industrial, research and educational workers and the crews of jet aircraft.

The number of such workers varies from country to country but a representative figure would be 1–2 per 1000, of which 0·3–0·5 per 1000 are medical workers. Occupations can be divided into high and low dose categories. The higher dose group includes industrial radiographers, radioluminizers, uranium miners, some nuclear reactor workers, nuclear fuel reprocessors, the crews of jet aircraft and the manufacturers of radiation sources for industry and medicine. The workers in these occupations receive annual average doses of 10 000–20 000 μGy, with a significant proportion of them receiving over 15 000 μGy (15 mGy).

The low dose group includes most nuclear power station workers, nuclear fuel manufacturers and most industrial and medical radiation workers. These workers are exposed to 1000–10 000 μGy, with very few workers receiving 15 000μGy.

The group of workers most likely to be over-exposed are industrial radiographers, while most workers in the nuclear power stations receive rather high doses and the highest average dose to a group of any size is to the nuclear fuel reprocessors at the British Nuclear Fuels plant at Windscale, recently renamed Netherfield.

Aircrew and cabin staff are subject to cosmic ray doses and assuming they spend 1 000 hours per year at 10 000–20 000 m altitude their average annual dose would be 2500–5000 μGy. Staff of supersonic planes flying for the same time at higher altitudes would receive 5000–15 000 μGy per year.

Industrial site radiographers have a poor safety record and some of the highest recorded individual doses. In the UK an accurate annual average dose is not available "but it is not greater than 10000 μGy" (UNSCEAR, 1977).

The average doses received by the manufacturers of radiopharmaceuticals such as those at Amersham International's radiochemical centre are rather high. The average annual dose decreased from 11 000 μGy in 1972 to 7600 μGy in 1974 but "a substantial proportion of workers received doses in excess of 15 000 μGy".

Most medical and industrial workers fall into the low dose category and receive annual doses of 1000–10 000 μGy with a few workers receiving over 15 000 μGy. For example, in 1974 in the UK approximately 6500 medical workers received an annual individual average dose of 2000 μGy.

For general industrial radiation workers there are few good statistics but in the UK a sample of about 10 per cent of (18 000) such workers showed that they received 4 300 μGy annual average dose. As we shall see in some detail in the next chapter some nuclear power workers are in the high and some in the low category.

The testing of nuclear weapons produces global contamination and the most important radio-isotopes of fall-out are those which give an external γ ray dose and those that become internally deposited in the body, either

Table 11.4. Dose commitments from nuclear explosions carried out before 1971

Source of radiation	Dose commitment (μGy) for the world's population			
	Lung	Gonads	Bone-lining cells	Bone marrow
External				
Short-lived	300	300	300	300
^{137}Cs	380	380	380	380
Internal				
^{3}H	20	20	20	20
^{14}C	90	70	290	320
^{90}Sr	–	–	710	520
^{106}Ru	240	–	–	–
^{137}Cs	170	170	170	170
^{144}Ce	380	–	–	–
^{239}Pu	9	–	9	–
Total	**1 600**	**940**	**1 900**	**1 700**

Source: UNSCEAR Report, 1977, Official Records of the General Assembly 32nd Session Supplement No. 40 (A/32/40).

by direct absorption or via a food chain. The method of fallout dose assessment is very complex because of the differential decay rates of the different isotopes and because of the influence of geographical and meteorological conditions on their deposition. UNSCEAR feels that the dose received per year is a figure of little value and that it is better to integrate the dose received from a given nuclear test or series of tests over a number of years. This dose is termed the total dose commitment of the population. Table 11.4 gives the dose commitment already received and to be received by the world's population by the year 2000 as a result of all the tests carried out up to the end of 1971.

Finally, there is a wide variety of miscellaneous sources of radiation to which the general population may be exposed. These may be classified into two categories. The first includes medical, military and industrial sources not normally available to the public but which find their way to the environment via transport accidents, theft, loss and incorrect disposal or misuse. Even though a few accidents have involved sizeable doses to a few individuals, the population dose from these sources is negligible compared with the natural background.

The second category includes such consumer products as radio-luminescent watches which used to contain radium-226. Alpha particles produced the fluorescence while β and γ rays gave rise to the exposure to the wearer. The radium content per watch varied widely (0·014–4·5 μCi; 5·2$\times$10^2–1·66$\times$10^5 Bq) and it has been estimated that a watch containing 1 μCi (3·7$\times$10^4 Bq) of radium-226 worn 16 hours per day would give an annual gonad dose in males of 150–300 μGy.

More recently tritium and promethium (soft β-emitters) have replaced radium and the genetically significant dose is considerably reduced since there is no external risk; the leakage of the isotopes from the watches might give the wearer an annual dose of 5–10 μGy.

The X-ray emission from a modern colour television is unlikely to exceed ICRP limits of 5 μGy h^{-1} at 5 cm; recent measurements give values of 1 μGy h^{-1} at 5 cm, giving an annual gonad dose of 2 μGy under normal viewing conditions.

The total annual average gonad dose from consumer products is probably less than 10 μGy and almost all of this is from radioluminous watches.

11.4. Summary of doses received by the population of the UK

The foregoing mass of detailed figures may be rather confusing and perhaps needs to be drawn together and simplified to enable us to gain a sense of proportion of the relative contribution of both natural and artificial sources. To do this we shall have to convert all the doses in tables 11.1–11.4 to units of dose equivalents in sievert. This will put all the different kinds of radiation on the same basis with respect to their ability to cause biological damage.

We saw in Chapter 1 that absorbed doses in gray are converted to units of dose equivalent sievert by multiplying the absorbed doses by the appropriate quality factor given in table 1.6. One sievert of any radiation will cause the same amount of damage to a given tissue. But some tissues are more susceptible to damage than others so we need yet another dose quantity! This quantity is called the 'effective dose equivalent' and is the dose equivalent weighted for the sensitivity to damage of different tissues. A list of the appropriate weighting factors and their derivation is given in the next section (p. 197). Table 11.5 gives a list of the annual average doses received by the population of the UK. The doses are all expressed as effective dose equivalents in sievert and they are taken from a 1981 publication of the National Radiological Protection Board (NRPB). The NRPB were set up by Parliament in 1970 to act as a "point of authoritative reference on radiological protection". From table 11.5 one can see the natural radiation dose (1 860 μSv) dominates all other

Table 11.5. Annual average dose to the United Kingdom population from various sources

Source	Annual effective dose equivalents (μSv yr^{-1})
Cosmic radiation	310
Terrestrial radiation	
External γ rays	380
Radon decay products	800
Other internal radiation	370
Total	**1 860** (~78%)[a]
Medical exposure	
Average dose	500 (~21%)
Genetically significant dose	120
Occupational[b]	
Average dose to workers	4 000
Average dose over whole population	9 (~0·4%)
Nuclear power industry discharge of radioactivity[b]	
Average dose to population	3 (0·1%)
Doses to most exposed individuals	1 000
Nuclear weapons testing fallout	
Average dose	10 (~0·4%)
Miscellaneous sources	
Average dose	8 (~0·4%)

Source: National Radiological Protection Board (UK), 1981.
[a]Figures in brackets represent approximate percentage contribution of each source to the overall dose to the population from all sources.
[b]For further details see Chapter 12.

doses with medical procedures (500 μSv) being the overwhelming artificial source of exposure for the population as a whole. All the other sources contribute very little with the exceptions of the small group of radiation workers in the population (average dose 4 000 μSv yr^{-1}) and an even smaller group of persons known as 'critical populations' (average dose 1 000 μSv yr^{-1}) that we shall discuss later in this chapter (see p. 200). The percentage contribution of each source to the overall effective dose equivalent of 2 400 μSv per year is as follows: natural background about 78 per cent; medical about 21 per cent; occupational, fallout and miscellaneous sources each with 0·4 per cent and discharges from the nuclear power industry 0·1 per cent.

The figures given in this section for natural and artificial radiation sources will form a useful guide when considering the risks of radiation. Some of the risk estimates have been touched upon in Chapters 7 and 9 and more will be mentioned in the next section which is concerned with the recommendations of the International Commission on Radiological Protection (ICRP).

11.5 Recommendations of the International Commission on Radiological Protection (ICRP)

The risks of excessive radiation were recognized almost immediately after Röntgen's discovery of X-rays in 1895 and precautions for the use of X-rays were quickly established in a number of countries. In 1928 the Second International Congress of Radiology founded the Commission, an international group of scientists, to keep the problems of radiation protection continually under surveillance. ICRP has published over 50 reports dealing with aspects of radiation protection and exposure and one of the recent ones, ICRP 26 (1977), is a general review of the most recent information in the field of radiation protection and contains recommendations for dose limits that are to supercede those of earlier reports of the Commission.

The ICRP stresses the importance of ensuring that all radiation exposure is justified and this involves the application of the three basic tenets of radiological protection. First, that the benefits of any acceptable radiation practice shall outweigh the detriment. Second, that in any such justifiable practice the radiation shall be kept '*as low as reasonably achievable*'—the so-called ALARA principle. Third, that only under very exceptional circumstances shall the dose limits recommended by the ICRP be exceeded. It is obvious that all these are profound statements containing words that demand definition and clarification and much has been written on the subject in an attempt to answer such questions as: How can one quantify and balance the risks and the benefits of radiation?

Who decides what constitutes a 'justifiable practice'? And what social, economic and practical considerations should apply to the ALARA principle? The answers and the value judgements attached to these questions take one outside the scope of this book and into the fields of social, political and even ethical judgements.

For radiation protection purposes the detrimental effects of radiation are called 'somatic' if they are expressed in the exposed individual and 'hereditary' if they affect his/her descendants. Somatic effects are recognized as being of two types:

(a) Non-stochastic effects in which the severity of the effect depends on the size of the radiation dose and for which there is a clear threshold of dose below which no detrimental effects are seen. Non-stochastic somatic effects include damage to blood vessels, the induction of cataracts of the lens of the eye and the impairment of fertility.

(b) Stochastic effects for which there is apparently no threshold dose and in which the severity of the damage is relatively independent of the size of the dose responsible for it. The predominant stochastic injury is the induction of cancer, where extremely low doses involve a finite risk of cancer. As the dose increases the severity of the individual malignancies does not increase, but there is a probability that a larger and larger proportion of an exposed population may develop the malignancy. The Commission assume that stochastic effects increase linearly with dose. Hereditary damage is also an example of a stochastic effect.

11.6. Estimation of risks

The induction of fatal cancer is now regarded as the predominant stochastic effect of radiation and this is a significant reversal for the ICRP which previously regarded hereditary effects as more important. The risk estimates for somatic effects made by the ICRP are based on epidemiological evidence of the sort given in Chapter 9 for cancer.

Table 11.6 gives the ICRP's estimated risk factors for fatal cancer induction and serious hereditary defects. The table also gives the relative weighting factors for the different organs of the body. These are based on the likelihood of induction of fatal cancer in the different organs, or in the case of the testes and ovaries the likely induction of hereditary damage. It is these weighting factors that are used in the determination of the effective dose equivalents (see p. 194).

The risk estimates for hereditary effects are derived from experimental mouse data (see Chapter 7). For purposes of radiation protection the ICRP gives an average risk factor of 4×10^{-3} Sv^{-1} for the damage expressed in the first two generations. This factor takes into account the proportion of radiation exposures likely to be genetically significant. The risk of hereditary damage from radiation including all subsequent generations is given as 8×10^{-3} Sv^{-1}. All risk factors in table 11.6 are very much average values and are intended to apply to an average individual irrespective of age or sex. For example, the risk of breast cancer in men is virtually nil but in women it is 1 in 200 per sievert (5×10^{-3} Sv^{-1}), so the 'average' value ascribed to breast tissue in table 11.6 is half the actual value for females and is given as 1 in 400 per sievert (2.5×10^{-3} Sv^{-1}). Red bone marrow is specified in table 11.6 because it is the tissue mainly involved in radiation-induced leukaemia. A risk estimate for radiation-induced leukaemia of 2×10^{-3} Sv^{-1} means that if a thousand people were each given a sievert then two of them would subsequently die of leukaemia. In addition to the specified tissues in table 11.6 there are 'all other tissues', for example, stomach, lower large intestine, salivary glands and probably liver, for which there is too little information for individual risk estimates to be made, although the total risk estimate is unlikely to exceed 5×10^{-3} Sv^{-1}. It is further assumed by the Commission that no single tissue is responsible for more than one-fifth of this value.

Table 11.6. Estimated risk factors for radiation induced fatal cancer and serious hereditary damage

Tissue/organ	Risk factor (Sv^{-1})	Fractional weighting factor
Testes, ovaries (hereditary risk)	4×10^{-3}	0·25
Red bone marrow	2×10^{-3}	0·12
Bone surfaces	5×10^{-4}	0·03
Lung	2×10^{-3}	0·12
Thyroid	5×10^{-4}	0·03
Breast	2.5×10^{-3}	0·15
All other tissues	5×10^{-3}	0·30
Whole body	1.65×10^{-2}	1·00

Source: ICRP Publication 26, 1977.

So, from table 11.6 one obtains a value for the overall risk per sievert for irradiation of the whole body and the fractional risk that each tissue contributes to the overall risk. This idea of fractional risk allows one to make some estimate of the risks from partial body radiation. Such non-uniform irradiation occurs in many practical situations, especially where radionuclides get into the body.

The total mortality risk factor for radiation induced cancer given by the ICRP is 1.25×10^{-2} Sv^{-1} or 1 in 80 per sievert for both sexes and all ages. This overall cancer mortality risk is thus much greater than the hereditary risk.

11.7. Dose equivalent limits (DELs) for occupationally exposed workers

The recommended occupational dose limits for whole body radiation is 50 mSv in a year, a limit which the ICRP feels ensures a degree of occupational safety consistent with that attained in other occupations not involving radiation and which are generally accepted as safe. This limit of 50 mSv is expected to prevent non-stochastic effects and to cause very few stochastic effects. Any recommendation of exposure limit necessarily involves judgements and comparisons of acceptable levels of harm and the Commission have compared risks in a number of occupations not involving radiation.

Table 11.7 gives a list of annual fatal accident rates in some UK industries and the estimated cancer risk among radiation workers. The latter is based upon the average effective dose equivalent received by UK workers given in table 11.5 as $4000\ \mu$Sv per year and the overall fatal

Table 11.7. Annual fatal accident rate in some UK industries and the estimated fatal cancer risk in radiation workers

Industry	Deaths per million workers per year
Deep sea fishing	2 500
Coal mining	250
Construction	200
Average radiation workers[a]	50
Textile	25
Clothing and footwear	3
All employment	50

[a]see text.

cancer risk of $1 \cdot 25 \times 10^{-2}$ per sievert. This implies an overall fatal cancer risk for the average UK radiation worker of $5 \times 10^{-5}\ \mathrm{yr}^{-1}$, or 50 deaths per million workers per year. This annual average risk of death is comparable with that in other 'safe' occupations. As we shall see in the next chapter some workers in some parts of the nuclear industry receive doses well above the average that must take their occupational risk out of the safest category. Indeed, if a significant fraction of the radiation workforce were to receive the recommended annual limit of 50 mSv over a number of years one can calculate that this would readily bring the radiation worker into the most hazardous occupation categories. In general however the ICRP believes that, while no occupational hazards are acceptable and all such hazards including radiation should be minimized, observance of the present dose limits (especially when the average dose workers receive is not more than one-tenth the maximum recommended) will ensure that the radiation industry ranks as a safe industry. The ICRP dose limit should be seen as an upper limit which workers should not ordinarily approach let alone exceed.

Although the Commission does not recommend dose equivalent limits for individual organs, 'implied limits' may be found by dividing the annual limit for the whole body by weighting factors which are given by the Commission. These factors are for individual organs and tissues and are derived by the ICRP from risk estimates for the induction of cancer and are given in table 11.6. Table 11.8 gives these implied dose equivalent limits (DELs) for different organs. It must be stressed that the Commission also recommends that to avoid non-stochastic effects no organ or tissue may receive an annual dose equivalent of 500 mSv (or 300 mSv for the lens of the eye), thus ensuring that the threshold for non-stochastic effects will not be exceeded even if such a dose were received every year for a working lifetime of 50 years. So the DEL for individual tissues or organs is whichever is the lower, the stochastic or the non-stochastic limit.

Table 11.8. Dose equivalent limits in mSv in a year

Tissue/organ	Radiation worker		Member of the public	
	Stochastic	Non-stochastic	Stochastic	Non-stochastic
Whole body	50	–	5	–
Gonads	200	500	20	50
Breast	330	500	33	50
Red bone marrow	417	500	42	50
Lung	417	500	42	50
Thyroid	1 670	500	167	50
Bone surfaces	1 670	500	167	50
Lens of the eye	300	–	30	–
Other single organs	833	500	83	50

Source: ICRP Publication 26, 1977 (data given or implied).

The Commission recommends that once women radiation workers are diagnosed as pregnant they can continue to work provided they are unlikely to receive three-tenths of the annual DEL. Prior to diagnosis of pregnancy the Commission consider it unlikely that the embryo would receive more than 5 mSv during the first two months a pregnancy might remain undetected.

There are exceptional circumstances where planned special exposures may be permitted provided the dose equivalent commitments from external exposure or intake of radioactive materials do not exceed twice the annual limit in a single event or five times the limit over a lifetime.

11.8. Recommended dose equivalent limits for the general public

The radiation protection of the general public involves two principles: (a) that no member of the public shall receive undue exposure, and (b) that the average exposure of the population shall be appropriately low.

The recommended maximum annual dose equivalent for any so-called 'critical group' of the population is 5 mSv to the whole body. This limit is exclusive of the natural background and deliberate medical exposure and is one-tenth of that for persons occupationally exposed. Such critical groups of people are often identified as receiving much larger doses than the average member of the population. The critical group may even be a single individual whose dietary predilection for fish eating may determine the discharge of radioactivity into the environment! Another important consideration is that the critical group may change with time and a nice example of this has occurred at the reprocessing plant at Windscale in the UK, now known officialy as Sellafield. Windscale's low-level effluent is discharged to the sea via a 2·5 km pipe and contains fission products and transuranic nuclides. The effluent used to be limited by the ruthenium-106 concentration in the edible seaweed *Porphyra umbilicus* that was collected on the Cumbrian coast and eaten by people in South Wales where it is known as laver bread. The seaweed trade has ceased and the Welsh get their laver bread from elsewhere. The critical group is now a few Cumbrian salmon fishermen who receive external γ radiation doses from the ^{95}Zr and ^{95}Nb in the sand and internal radiation doses from ^{137}Cs and ^{134}Cs concentrated in plaice, dab and skate.

The Commission makes a number of recommendations for reviewing the different routes and practices which may result in the exposure of the general public. If such precautions are taken it is believed that the lifetime dose equivalent to any member of the public would not exceed 1 mSv per year.

The Commission no longer proposes dose limits for the population as a whole; instead it believes that the application of dose equivalent limits for individuals as well as an observance of its general principles will ensure that the average dose equivalent to populations will not exceed 0·5 mSv per year. Using the overall risk factor for fatal cancer derived from table 9.4 as 1 in 80 per sievert this 0·5 mSv dose implies an average fatal cancer risk of 1 in 160 000 per year. Table 11.9 gives the annual average probability of death from some activities, some voluntary, some not so voluntary. It serves merely to give some perspective of mortality risks, the figures given are only approximate to an order of magnitude.

One interesting and potentially important recent finding that may have direct relevance to radiological protection is that some people may be abnormally radiosensitive. In Chapter 2 we noted that several genetic diseases (Bloom's syndrome, Fanconi's anaemia, Progeria, Ataxia

Table 11.9. Annual probability of death for an individual

Risk	Causal factor
1 in 200	Smoking 20 cigarettes per day
1 in 6 000	Traffic accidents
1 in 10 000	Domestic accidents
1 in 20 000	Industrial accidents
1 in 30 000	Drowning
1 in 100 000	Poisoning
1 in 1 000 000	20 minutes being a man of 60 years old; Driving 50 miles by car; 90 seconds rock climbing
1 in 2 000 000	Struck by lightning

telangiectasia, Cockayne's syndrome, etc.) are characterized by defective DNA repair. These patients are often highly sensitive to UV or X-radiation and are cancer prone. It has been estimated that 1–5 per cent of the human population may be heterozygous for mutations that interfere with DNA repair. Such a predisposition to radiosensitivity cannot be detected at present and so the consequences for radiological protection have yet to be considered.

It is obvious that the acceptance of risks is a complex phenomenon and is far outside the scope of this book. Suffice it to say that the acceptance of one risk is no justification for the acceptance of another and it is irrational to expect people to forget small risks (such as those associated with nuclear power production) simply because they are inevitably exposed to larger risks. On the other hand, it is equally irrational of opponents of nuclear power to distort or magnify these small risks without putting them into context. It is often said that human beings are the prisoners of their own prejudices and it behoves all protagonists in the nuclear debate to examine all the data and to remain sceptical of their own and their opponents convictions.

11.9. Radiation protection and radionuclides

The recommended dose equivalent limits in table 11.8 are for radiation from all sources, whether internal or external to the body. The external hazards from radionuclides come largely from those that emit γ rays since the α and β particles emitted by most nuclides are generally not energetic enough to penetrate deeper than the most superficial layers of the skin. In contrast, in the case of internally deposited radionuclides, the major hazard comes from those nuclides that emit α and β particles because all their energy is dissipated locally in short densely ionizing tracks. A further hazard from internally deposited radionuclides is that they may be selectively accumulated in certain 'critical' organs of the body, for example, calcium-45 and strontium-90 in bone and iodine-131 in the thyroid. The assessment of the relative toxicity of different radionuclides inside the body is therefore complex and involves detailed information on the following points:

(a) the energy and type of radiation emitted by the radionuclides;

(b) the physical half-life of the radionuclide, i.e. the decay rate;

(c) the biological half-life, i.e. the time for the elimination of 50 per cent of the atoms of the radionuclide from the body; and

(d) whether localized concentration of the radionuclide occurs in critical organs of the body.

The accurate determination of the biological factors (c) and (d) is often very difficult and dependent upon numerous other factors including both the physical nature (gaseous, liquid, solid, particle size, etc.) and the chemical nature of any compound that contains the radionuclide and in particular the so-called 'biological solubility'. It is also necessary to know not only what fraction of a radionuclide is actually incorporated in the body but into what organs it goes and at what concentration and the subsequent translocation (if any) to other organs. Coupled with this one needs a detailed knowledge of the excretion rate from the different organs and the overall pattern of excretion from the body. From such considerations the biological half-life can be determined and a combination of the physical and the biological half-lives gives one the all important 'effective half-life'.

A special committee of ICRP gives guidance on internal radiation hazards. Such guidance used to involve concepts of 'maximum permissible body burden' (MPBB) and 'maximum permissible concentration' (MPC) and although these have now been replaced they are still in use and are defined as follows:

MPBB—the limit for a particular radio-isotope in the body of a radiation worker that would cause him to be irradiated to the level of the basic recommended standards.

MPC—the limit for a particular isotope in air or drinking water that for a radiation worker exposed to such concentrations for 40 hours per week would cause him to be irradiated to the level of the basic recommended standards.

We noted above that the total dose an organ will receive from a given intake of radionuclide is a complex function of its rate of build up in the organ coupled with the radioactive decay and the biological elimination/excretion of the material.

This total dose integrated over a working lifetime of 50 years is known as the committed dose equivalent (H_{50}) and from it the ICRP now establishes what are called annual limits of intake (ALI) for different nuclides. The ALI is the amount of radionuclide in becquerel (Bq) that would give a harm commitment to the organs it irradiates equal to that from whole body irradiation of 50 mSv per year.

The hazards of internal irradiation are controlled by limits recommended for airborne and work surface contamination. These are called 'derived air concentrations' (DACs) and 'derived working limits' (DWLs) and have been calculated for a wide range of radionuclides.

11.10. Radiation protection in practice

In order to comply with the ICRP's recommended dose levels it is essential that personnel concerned with radiation be continuously monitored so as to record any dose that they receive during their work. There are two types of personnel monitoring devices in common use for external radiation, the film badge and the pocket ionization chamber dosimeter.

Film badges are worn by personnel and the accumulated dose the film receives over, for instance, 14 days is determined. By using filters of various thickness and material to cover parts of the film it is possible to get an idea of the quality of the radiation to which the film has been exposed, for example, 'soft' (long-wavelength) X-rays will not penetrate a thick filter. The main advantage of the film badge is its simplicity, and its main disadvantage is that the personnel know only the total dose they have received over the past 14 days. If they have received an overdose they have no real check as to where it was or when it was received.

Pocket dosimeters are small ionization chambers, and their great advantage is that they give an immediate and precise indication of exposure dose. This immediacy allows the worker to pinpoint the source of the radiation and to take steps to avoid future exposure. Pocket dosimeters are invaluable for personnel going into known radiation areas. Occasionally thermoluminescent devices (TLDs) are also used, for example, when handling some radioactive sources small finger sachets of thermoluminescent material are advisable.

To comply with the ICRP's recommendations that all doses be kept as low as practicable or achievable, there are three safety precautions that can be taken to ensure minimal exposure to external sources of radiation, they are:

1. Any radiation *exposure time* is to be kept to an absolute minimum.
2. Since the intensity of electromagnetic radiation falls off as the square of the *distance*, radiation workers should keep as far away from sources as is practicable (for instance, if a radiation source gives 100 Gy h^{-1} at 10 cm, it will give 1 Gy h^{-1} at 1 m).
3. If the radiation source is intense, *shielding* with such materials as lead may be necessary. The thickness of material required is obtained by using half value thickness calculations (see Chapter 1).

Besides these safety precautions there are detailed procedures and laboratory rules to minimize the external and internal hazards involved in handling radio-isotopes. The external hazards from the small amounts of isotopes used as tracers ($<$ 1 million Bq) are usually low. In contrast, internal contamination from isotopes may pose a health hazard even if very small amounts are being handled. The exact details of the regulations will vary with the quantities and the hazard classes of the isotopes being handled, but their object is to avoid contamination of the laboratory, its atmosphere, its equipment and its personnel. Thus eating, drinking, smoking, applying cosmetics and pipetting by mouth are forbidden in isotope laboratories. The wearing of special clothing (rubber gloves, overalls and overshoes) is sometimes necessary and any spills of radioactive material should be monitored and reported to the radiation protection officer. The continuous monitoring of the air in special laboratories may be necessary. All radioactive material should be rigorously labelled and the disposal of solid, liquid and gaseous waste should be strictly monitored and in accordance with permissible levels. At the end of any experiment, equipment and personnel may have to be monitored.

There are essentially two objectives when monitoring workers for internal contamination. These are to determine qualitatively whether an exposure has occurred and to determine quantitatively organ or body burdens. Essentially, internal monitoring is an adjunct to external surveys and is routinely required at, for example, facilities reprocessing plutonium and transuranics. The frequency of such measurements depends on (a) effective half-life, (b) variation of excretion rate and (c) recent contamination experience. For materials with long effective half-lives, for example, plutonium, the interval between monitoring should be chosen to assess the long-term build up; *quarterly* measurements for those with highest potential risk; *annual* measurements for those with minimal potential exposure.

There are three methods in common practice: bioassay, whole body counting, and surveying the working environment. Bioassay is the analysis of body fluids, excreta and tissues for their radioactivity. The type of assay varies with the radio-isotope under assay. The procedures are most useful where the radioactive substances are 'soluble', i.e. rapidly transferred to body fluids. Inhaled insoluble material, for example, plutonium oxide particles, are likely to be retained in the lung and not detected in urine or blood. However, particles will in part be swallowed and may appear in the faeces. The analysis of urine, urinalysis, is the commonest routine method to estimate the body content of radio-nuclides. Allowance must be made for the natural ^{40}K content of the urine and the sensitivity of a system must be sufficient to detect small fractions of maximum permissible body burdens. The body burden is estimated using metabolic models of retention and excretion of varying complexity. Usually knowledge of the time of exposure and a series of samples are required before valid estimates of the initial body burden are possible. However, the accuracy of bioassays is not great due to inherent inaccuracies in the biological models, individual variations and the process of translating bioassay data into estimates of body/organ burdens. Blood analysis gives very similar information for total body burdens but it is rarely used except for suspected acute intakes of soluble material.

Finally, sputum, nasal smears, throat swabs and faeces can be used to determine inhaled or ingested material. Airborne particles (especially those with a mean aerodynamic diameter greater than 10 μm) will be trapped in the upper airways of the lung and carried to the throat and swallowed. Much of this, if it is insoluble, will appear in the faeces. These assays are useful in surveys of accidents or to check previously unsuspected chronic contamination.

If the radionuclides taken into the body emit γ rays or an energetic β particle they can be detected by external partial or whole body counting equipment. Such methods have an accuracy of ± 30 per cent, i.e. much more accurate than most bioassay methods.

It is often said that radionuclides should be treated as if they were pathogenic bacteria. it is a good maxim.

Chapter 12
Nuclear power and the environment

12.1. Introduction

Nuclear power and the radiobiological hazards associated with the nuclear fuel cycle are among the most controversial of current public debates. The aim of this chapter is to outline the more important aspects of the public health and environmental effects of the operations involved in nuclear power production.

The Wykeham Series Book No. 51, *Nuclear Power and the Environment* by R. J. Pentreath, gives a detailed account of many of the topics covered in this chapter and it is to be recommended. Here we shall first outline the stages of the nuclear fuel cycle, then look at the hazards associated with each stage and the doses received by radiation workers at each stage, and finally we shall make some general comments on the so-called 'nuclear debate'.

12.2. The nuclear fuel cycle

Most of the world's reactors are called thermal reactors and they are fuelled by uranium. There are very few fast reactors fuelled with a mixture of plutonium and uranium.

The fission of uranium into fragments (see p. 5) is accompanied by a massive release of energy, mostly as heat, which is converted to steam that drives electricity generators. Figure 12.1 shows the main features of the nuclear fuel cycle from the mining of uranium to the reprocessing of the irradiated fuel from the reactor. Uranium ores, containing about 0·2 per cent uranium with its associated decay products thorium and radium, are mined throughout the world. The uranium is separated from the ore as oxides and for most reactors the naturally more abundant ^{238}U must be

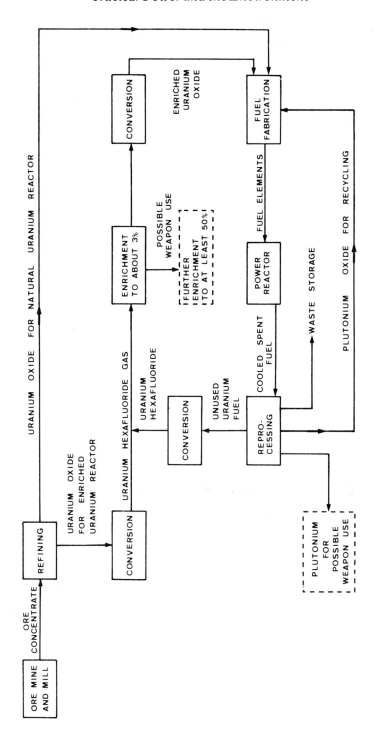

Figure 12.1. The fuel cycle

enriched with [235]U. The proportion of the latter must be increased from its natural value of about 0·7 per cent to 2–4 per cent in order to enhance the fuel's ability to sustain a chain reaction.

The enriched uranium hexafluoride is chemically processed to the metal or to the oxide and then made into fuel 'elements' which are clàd in gas-tight metal tubing. New fuel elements are transported to the power stations where they replace the irradiated or 'spent' elements. The loading of elements into a reactor may be done while it is running or it may involve a 'shutdown'.

The spent elements are intensely radioactive as a result of the fission process. Radioactive materials are also produced in the reactor by the interaction of slow neutrons with structural materials, corrosion products and cladding. So, before being transported for reprocessing the irradiated elements spend a few weeks in specially designed 'cooling ponds'. This allows the shorter lived isotopes to decay while at the same time the fuel rods are thermally cooled and shielded by the mass of water in the pond. A nuclear reprocessing plant (such as that at Netherfield, England) recovers the valuable fission products and obtains even more valuable actinides such as plutonium, americium and curium. Plutonium is especially important since it is used in bomb making, is a potential substitute for [235]U in the enrichment of fuel in thermal reactors (1 per cent plutonium would be needed) and is the fuel of fast breeder reactors, where 30 per cent plutonium is needed. Over 99·9 per cent of the radioactivity produced in fuel elements is locked up within the elements until they are broken open for reprocessing and eventual recycling. Before we look at the potential hazards associated with the stages just described we shall briefly outline the general design of nuclear reactors.

12.3. The general design of a thermal nuclear reactor

There are a number of different types of reactor but they all consist of:
1. a core of fuel elements, a moderator and control rods;
2. cooling circuits, coolant and heat exchangers;
3. containment vessels and buildings; and
4. a multiplicity of monitoring and control systems that continually check all aspects of the state of the reactor.

Figure 12.2 is a diagram of the American-designed pressurized water reactor (PWR). The reactor consists of a steel pressure vessel filled with water at 320°C under 150 atmospheres pressure (15 MN m^{-2}) inside which is the reactor core. The core consists of about 200 fuel elements together with the control rods. The pressurized water acts as both the moderator and the coolant. The water passes in a closed circuit through the pressurizer via the core to the heat exchanger. The design also

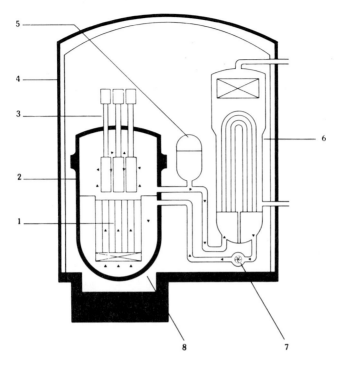

1 Core
2 Steel pressure vessel
3 Control rods
4 Steel lined concrete shield
5 Pressuriser
6 Steam generator
7 Primary coolant pump
8 Sump

Figure 12.2 Diagram of a cross-section of a pressurized water reactor (PWR)

incorporates a number of emergency core cooling systems and the entire pressure vessel is further contained inside a massive concrete or pressure containment building.

The function of the moderator is to slow down the fast neutrons produced in the fission process and so enhance their chance of reacting and splitting other uranium nuclei. The function of the control rods is to absorb excess neutrons that are produced during the fission process. By withdrawing or introducing the rods the speed of the nuclear reaction can be adjusted to maintain 'criticality'. A controlled chain reaction is critical when each neutron that is lost by causing a fission is replaced by one more.

When a reactor is loaded with fresh fuel all the control rods are in the core and they are withdrawn gradually and in a symmetrical manner so as to maintain a constant neutron flux throughout the core. This is done until criticality is reached. The reactor is then run in this condition until the fission products build up in the fuel, some of which, for example, ^{135}Xe, slow down the system by absorbing neutrons. This, together with the gradual exhaustion of the fuel, means that as the reactor ages more control rods need to be withdrawn from the core to maintain criticality. Eventually there are insufficient neutrons for further fission and new fuel elements are required.

12.4. Hazards associated with the nuclear fuel cycle

Mining

Due to its extremely long half-life (see Appendix, p. 229) naturally occurring uranium is only slightly radioactive, but its decay products such as thorium, radium and particularly radon do present a radiation hazard to miners. Radon-222 itself contributes only 5 per cent of the α radiation miners are exposed to, the main hazard coming from the decay daughters of radon: ^{214}Po, ^{218}Po, ^{214}Pb and ^{214}Bi. The atmosphere of the mines contains inert dust particles to which these and other radionuclides are attached and their inhalation can cause high levels of lung cancer (see Chapter 9). In modern mining the radon concentration is controlled by ventilation, the air is filtered and respirators are used to limit radiation exposure.

Milling

The uranium ore is crushed to the texture of fine sand from which the uranium is leached. This is followed by a complex series of treatments that concentrate the uranium, mainly as oxides. This series of processes is called 'milling' and the great bulk of the initial ore that is left is known as 'tailings' (equivalent to the slag heaps at coal mines).

These tailings contain significant concentrations of radon and radon daughters. The US Environmental Protection Agency has recently expressed concern that ^{230}Th, ^{226}Ra and ^{210}Po can and do enter the atmosphere and water pathways from tailing piles. There is also concern about the internal doses received by individuals from the inhalation of radon daughters from the tailings that have to be incorporated into building materials in places like Grand Junction, Colorado, USA.

Enrichment

The milling process produces relatively pure oxides but these must be further purified, usually by dissolution in nitric acid. The uranyl nitrate produced is converted to uranium hexafluoride (UF_6) which is a gas at temperatures above 58°C and which is ideal for either the gaseous diffusion or the gas centrifugation enrichment processes. The enrichment of the fuel increases the proportion of ^{235}U from 0·7 per cent (its natural concentration) to 2–4 per cent. The 'depleted' uranium left behind containing about 0·3 per cent ^{235}U is the waste product of enrichment. Finally, the enriched UF_6 is converted to uranium dioxide (UO_2) which is the fuel of most thermal reactors.

The main hazards to workers at enrichment plants are not radiological but are the extreme toxicity and corrosiveness of fluorine, hydrofluoric acid and uranium hexafluoride.

Fuel fabrication

The manufacture of enriched uranium fuel elements involves no special radiological precautions except the control of criticality. This entails very strict limits on storage location and the prevention of water entering areas where it might act as a moderator and so allow fission to occur. All fuel elements have to be very precisely made as they must remain dimensionally stable, despite the extreme stresses during irradiation in the reactor, to avoid the escape of fission products. The elements are encased in such metals as stainless steel, magnesium alloy (magnox) or zirconium alloy (zircalloy). The manufacture of plutonium fuel for fast reactors is a much more hazardous and difficult procedure.

Transport

The new fuel is transported by ordinary means to nuclear power stations. The containers have to conform to the stringent standards of the International Atomic Energy Agency (IAEA) and with those of the country in which they are travelling.

The 1976 Report of the Royal Commission on Environmental Pollution said that present transport practices in the UK did not appear to be a significant public hazard at present, but this might not be so in future, when such traffic may be greater and involve active cooling of the fuel in transit.

So far we have seen that the radiation hazards from uranium mining, milling, fuel enrichment, fabrication and transportation are considered relatively minor. The more serious problems arise in power stations and more particularly at reprocessing plants which deal with the highly active fission products of the spent fuel elements. We shall deal with power stations and reprocessing plants separately.

12.5. Nuclear power plants and the management of radioactive waste

In the reactor the fuel undergoes fission and energy is released as heat which converts water to steam that drives the turbines that produce electricity. In a large power station (500 MW(e)) the radioactive fission products amount to a few kilograms per day and 99·9 per cent of the activity remains confined inside the metal cladding of the fuel elements. Besides fission products, structural materials of the reactor and the components of the cooling system corrode with time. These corrosion products, commonly iron, manganese and cobalt, together with impurities in the coolant, are activated by neutrons in the reactor and become radioactive. The amounts of such material are very small compared with fission products.

In all reactors fuelled by uranium the radioactive wastes are basically the same although important differences in effluents do occur which are related to the reactor coolant and the steam cycles used. Nuclear power stations are designed to minimize both the volume and radioactivity of their gaseous, liquid and solid effluents and must, by continuous monitoring, ensure that no release exceeds authorized permissible levels. Let us look briefly at these three types of waste and how they are managed.

Gaseous waste from nuclear power stations

The main gaseous radionuclides released from nuclear power stations include the noble gases (^{41}Ar, ^{85}Kr and ^{133}Xe), ^{129}I and ^{131}I, ^{14}C, ^{3}H and ^{13}N. Most of these are of little environmental significance either because they occur in such small quantities or have such short half-lives (see Appendix, p. 229).

^{131}I is a volatile fission product which escapes by leakage from fuel elements, and would be one of the principal sources of danger to the public in the event of an accidental release since by direct inhalation or ingestion in cow's milk it can irradiate the thyroid gland.

^{14}C is produced in reactors that use carbon contaminated with nitrogen as a moderator. The amount released throughout the world is much less than 1 per cent of the naturally occurring ^{14}C in the atmosphere.

^{3}H is produced in reactors using heavy water (D_2O) as the moderator, by neutron capture by deuterium.

The waste management methods used for gaseous wastes are threefold: (*a*) delay and decay, (*b*) filtration and (*c*) adsorption on activated charcoal. Delay and decay involves allowing radioactive decay to occur, thus reducing the need for any other treatment. Filtration of gases collects suspended radioactive particles produced when a gaseous parent nuclide decays to a particulate radioactive daughter. Activated charcoal filters are used for example for idodine, and are useful in the delay and decay of the noble gases. All airborne radioactivity is continually monitored and at present the discharges do not present a significant hazard.

Liquid waste from nuclear power stations

Nuclear power stations produce relatively small quantities of liquid radioactive wastes and these are mostly found in the cooling ponds where the spent fuel elements are stored for three months or so to allow short-lived fission products to decay. The ponds also contain the neutron activated corrosion products from the metal cladding of the fuel elements, typically ^{55}Fe, ^{51}Cr, ^{54}Mn and ^{60}Co, plus small amounts of ^{3}H. Delay and decay in waste 'hold-up' tanks, filtration, evaporation and ion exchange resins (demineralizers) are all used to concentrate and alter the composition of the wastes. This wet sludge can be measured for its radioactivity, re-used in the plant, stored or discharged. Discharges are always monitored for control and public protection purposes. They are released into rivers, lakes and the sea at rates determined by the capacity of the environment to disperse and dilute the effluent. Biological reconcentration pathways and routes back to 'critical' populations of man must be considered. For example, at Bradwell nuclear power station on the estuary of the Blackwater in Essex there is an oyster hatchery. Oysters can concentrate zinc and silver from seawater by factors of 10^5 or more and this effectively limits the levels of ^{65}Zn and ^{100}Ag that can be discharged by the power station (see also p. 200).

Solid wastes from nuclear power stations

The solid radioactive waste produced in power stations includes parts from the core, control rods, neutron flux measuring instruments, filters, valves, ash from incinerators, sludges and used ion exchange resins. Most of this accumulates 'on site' and is stored in concrete vaults. In the United Kingdom each power station is expected to accumulate about $2\,400\,m^3$ of solid waste over its 25 year lifespan. Most of it will decay to harmless levels in about 100 years.

12.6. *Fuel reprocessing and the management of radioactive wastes*

Fission products accumulate in the fuel elements and by absorbing neutrons they eventually interfere with the fission process. Such spent elements have to be removed from the reactor and after a period in the cooling ponds of the power station they are transported to reprocessing plants. The purpose of reprocessing is the recovery of the plutonium produced in the reactor and of the unburned uranium which can be re-used.

The fuel elements are disassembled by remote control methods, their metal cladding removed and the fuel sheared into small pieces and dissolved in boiling acids. The plutonium and uranium are extracted by a series of solvent extractions and ion exchange treatments. More than 99 per cent of the fission products appear in the aqueous phase. At all stages of reprocessing large quantities of gaseous, liquid and solid wastes arise and create waste management problems. The crucial distinguishing feature of radioactive waste in reprocessing plants is that much of it is contaminated with plutonium and other transuranic elements which are very long lived α-emitting radionuclides that must be permanently isolated from the environment.

Gaseous wastes from reprocessing

The gaseous wastes from reprocessing come mainly from fission gases and volatiles, ^{85}Kr, ^{3}H and ^{129}I being of particular importance. There are also radioactive dust particles and suspended fine spray from radioactive liquids and nitric acid fumes. The gases are routinely cleaned before monitoring and release via high chimneys. The cleaning processes include inertia demisting, filtration, electrostatic precipitation and 'scrubbing' (absorption with water, nitric acid or alkali). These methods remove everything except the ^{85}Kr, ^{3}H and ^{129}I, and they are monitored and released into the atmosphere.

Liquid wastes from reprocessing

Liquid wastes are divided into low, medium and high activity wastes, the latter containing over 99·9 per cent of the fission products.

Low level liquid wastes are normally released into the rivers or the sea under carefully controlled conditions. The permitted releases are established by a 'critical path approach'. Surveys are carried out by the authorizing ministries to identify all the ecological routes by which critical radionuclides may be reconcentrated and returned via food chains and other pathways to critical groups of people. Examples of such critical pathways include the concentration of ^{137}Cs in fish, [106]Ru in seaweed, [65]Zn in shellfish and [95]Zr adsorption in sand to which fishermen may be exposed.

Medium level liquid wastes in reprocessing plants are variously treated (evaporation, decay storage, and chemical precipitation) so as to produce dispersable low level wastes and a little high level waste. Some intermediate waste is isolated and secured in concrete silos.

The wastes we have dealt with so far contain less than 1 per cent of the long lived radioactivity produced by the nuclear power industry. These wastes can be deliberately and safely released in a planned and monitored way into the environment. However, there is a need for continual vigilance by the authorities since levels of discharges tend to rise year by year. For example, the release of ^{137}Cs from Sellafield into the Irish Sea has been excessive in recent years although it does not cause public exposure in excess of the ICRP dose limits. However, it is the high level liquid waste containing over 99 per cent of the fission products that poses the most intractable problem. This waste is an acid solution containing most of the elements from arsenic to europium plus α-emitting actinides produced from uranium by successive neutron capture. The most important actinides involved are isotopes of plutonium, americium and curium. After extraction of much of the plutonium and uranium, the liquids are allowed to boil under their own radioactive heat, so concentrating the liquor. The intensely active liquids will have to be stored for hundreds or even thousands of years because of their long half-lives and extreme radiotoxicity if taken into the body.

The high level liquid wastes are stored in exceedingly complex double skinned stainless steel and concrete storage tanks with several independent cooling systems to remove decay heat. Air is blown through the liquid to keep precipitated solids in suspension and gaseous effluent is cleaned on scrubbers and electrostatic precipitators. Removal of decay heat will be necessary for several hundred years; if the solutions were to boil dry the decay heat would cause serious and widespread release of volatile materials. The activity of radionuclides such as ^{90}Sr means total containment for hundreds of years and if there are considerable concentrations of transuranic elements, such as [239]Pu, containment and surveillance will have to be perpetual.

The volume of such high level liquid wastes stored at Sellafield in 1976 stood at about 650 m³, with a total activity of about $1 \cdot 5 \times 10^{19}$ Bq, and the volume expected by the year 2000 is over 6000 m³ (about 5×10^{20} Bq). At present the authorities consider such tank storage to be safe provided there is both sufficient surveillance and spare tank capacity to accept liquids from any leaking tanks. The most recent advice by the UK Radioactive Waste Management Advisory Committee (1981) is that high level waste should be stored in sub-surface or above-ground containers for at least 50 years and possibly much longer prior to deciding its final resting place—deep underground, on the ocean bed or under the ocean floor. There are plans to mix the liquid waste with silicates and borates and to dry, calcine and eventually vitrify the mixture. There is much current research into the possibilities of more or less completely separating the long-lived actinides prior to vitrification so that the secure storage of the glass blocks only lasts a few hundred years and not hundreds of thousands of years. If the vitrification and actinide separation problems can be solved there only remains the problem of the ultimate, irretrievable and permanent disposal of the glass blocks. They could be stored on land provided they were appropriately maintained and guarded. Such perpetual uninterrupted supervision is hardly feasible—apart from the possibility of natural and human intervention that might disperse the material, such storage puts an inordinate responsibility on future generations. Therefore, two disposal options have been proposed: (1) deep geological disposal in dry stable formations on land and (2) disposal on the deep seabed or, more likely, in holes in the deep seabed that are then backfilled. The former is perhaps the more likely to be adopted. However each of the proposals is fraught with difficulties. For example, it will be necessary to find a geological site not prone to seismic events, containing only minerals that will never be of commercial value, absolutely inaccessible to groundwater and having favourable 'thermal conductive properties'. A lot of mathematical modelling of the characteristics of such a site has been done but there is a lot more practical research to do before this disposal option becomes a reality.

The problems of undersea disposal are even greater. These include the sheer technical difficulties of burial; again there is the need to be certain of the geological stability of the site, and there would be inevitable corrosion and leaching by water in the sediments. Coupled with these problems is an almost total ignorance of the physical, chemical and biological characteristics of the deep oceans. Finally, there is the international political consideration of who owns the seabed anyway! Other facile suggestions for disposal include Antarctica or the use of rockets to dispose of it into the sun or deep space.

In conclusion, it would seem that as a result of massive engineering and scientific effort the waste management techniques now available have so far proved effective. However, a pertinent criticism of the ever-increasing commitment to nuclear power hinges on the need for the nuclear authorities to show "beyond reasonable doubt" that they can safely contain large quantities of radioactive waste for thousands of years. The nuclear industry is confident that its ingenuity will enable it to cope with ever-increasing volumes of waste. Its critics have less faith.

12.7. Doses received by workers at various stages of the nuclear fuel cycle

In this section we shall summarize the occupational doses that are known to be received by workers at the various stages of the fuel cycle, from the uranium miners to the nuclear fuel reprocessors. The doses will be given in two ways: as average annual doses received by individual workers and as annual collective doses. The annual collective dose to a population is defined as the weighted sum of the individual doses received in one year, multiplied by the number of people exposed. This is expressed in units of man-gray. Such collective doses can be used to compare the relative risks incurred by groups of differing sizes exposed to different sources and types of radiation. it also takes account of the fact that the actual doses to the individuals will vary with time. As we shall see the parts of the fuel cycle which give the highest collective doses are the reactors, fuel reprocessing and research and development. Mining, milling, fabrication and transportation involve smaller collective whole body doses.

Uranium miners

The dose received by uranium miners involves internal irradiation from the inhalation of radon and its daughters and external whole body radiation.

Exposure to radon and its daughters in the air of the mine is defined as the integral of the activity concentration in the air over a certain exposure time. The unit most often used is the working level month (WLM) which is the exposure during 170 working hours in a radon daughter concentration of 1 working level (WL) and this corresponds to an equilibrium concentration of $1 \cdot 7 \times 10^4$ pCi l^{-1} ($6 \cdot 3 \times 10^2$ Bq l^{-1}).

The radon and thoron daughters are often attached to aerosol particles in the atmosphere and the deposition of these particles in the respiratory system of miners is very complex. The particles can reach deep into the lung whereas the more soluble, unattached radon and thoron daughters are deposited in upper airways. The dose to the lung is therefore very inhomogeneous and is influenced by the rate, depth and route (nose or mouth) of respiration, the geometry of the airways and the clearance and translocation patterns in the lungs.

A WLM delivers about 10 mGy of α radiation to the bronchial epithelium and an estimated 5 mGy to the whole lung.

Uranium miners used to be exposed to very high levels of radiation and up until the 1960s this was the main radiological problem associated with the fuel cycle. Indeed, prior to the ventilation of the mines and the use of respirators miners in the last century and first few decades of this century were very likely to die of radiation-induced lung cancer. For example, in the silver, copper and uranium mines at Schneeberg and Joachimsthal in Czechoslovakia in the first few decades of this century miners were exposed annually to between 240 and 360 WLM. Between 1930 and 1950 42 per cent of the miners at Schneeberg and 68 per cent of those at Joachimsthal died of lung cancer, the majority of it due to radiation exposure during their working life in the 1910s and 1920s.

However, there have now been very marked improvements in the ventilation and conditions in the mines so that in 1974 the average annual exposure of United States miners was 1·9 WLM and of French miners, 1·3 WLM. These levels correspond to annual bronchial doses of 15–20 mGy. The average annual external dose for most uranium miners throughout the world is estimated to be 10 mGy.

Milling and fuel fabrication workers

The annual average dose in the milling and fabrication steps of the fuel cycle are small. The average individual doses and the annual collective dose for UK workers are given in Table 12.1.

Table 12.1. Occupational doses for UK fuel enrichment and fabrication workers in 1975

	Annual average dose (μGy)	No. of workers	Annual collective dose (man-mGy)
Fuel manufacture:			
Chemical processes	4 500	220	900
Fabrication processes	5 800	120	700
Canning and assembly	3 300	61	200
Fuel maintenance	3 500	500	1 750

Source: UNSCEAR Report, 1977.

The 1977 UNSCEAR Report notes that the likely future increases in plutonium fuel fabrication for fast reactors will increase the potential for plutonium intake.

Fuel reprocessors

These workers have average doses among the highest reported for any industrial workers and there are a significant number exposed to dose equivalents in excess of 15 mSv per year. Table 12.2 shows the annual average occupational doses and the annual collective dose of the Sellafield (Windscale) fuel reprocessors. The doses are given in millisieverts, mSv.

Table 12.2. Distribution of dose for fuel reprocessing workers at Windscale (Sellafield), UK, 1975

Dose range (mSv)	Number of workers
⟨5	1 603
5–10	507
10–15	283
15–50	952
⟩50	36
Total	**3 381**

Average annual dose 11·9 mSv
Annual collective dose 40 280 man-mSv

Source: UNSCEAR Report, 1977.

Workers at nuclear power reactors

There is a mass of information throughout the world on the occupational doses of workers in all types of thermal nuclear reactors. As an illustration we shall use UK data as typical. Table 12.3 shows a summary of the occupational doses for 1974 for workers at a number of commercial reactors.

It can be seen that the doses at the newer stations, Oldbury, Sizewell and Wylfa, are lower than at the older ones, Berkeley, Hinkley Point and Hunstanton.

Table 12.3. Summary of occupational doses at UK gas cooled power plants in 1974

Plant	Average annual dose (mGy)	Annual collective dose (man-Gy)
Berkeley	7	2 840
Bradwell	3·2	1 290
Hinkley Point	3·1	5 140
Trawsfynydd	4·3	2 600
Dungeness	2·0	1 350
Sizewell	1·6	820
Oldbury	1·7	710
Wylfa	1·2	720
Hunstanton	5·0	3 600

Source: UNSCEAR Report, 1977.

Transport of nuclear fuel

The transport of unirradiated fuel, spent fuel and solid wastes leads to small doses to the drivers of the order of 0·003 mSv per shipment.

Research and development

A final class of worker must be included in any estimate of the total collective dose received by people concerned with the generation of nuclear electricity. This is the research and development workers of the various national atomic energy authorities. The range of doses received by such people at United Kingdom Atomic Energy Establishments between 1972 and 1974 are given in table 12.4, the dotted line shows the recommended annual dose equivalent limit of 50 mSv. Using data of the type given in table 12.4, it can be calculated that in 1974 the annual collective dose was 39·6 man-sieverts, the annual individual dose being 5·7 mSv.

Table 12.4. Dose received by workers at the UK Atomic Energy Establishments 1972–74

Dose range (mSv)	Number of workers		
	1972	1973	1974
⟨15	5 949	5 798	6 136
15–30	694	648	587
30–40	258	189	142
40–50	152	115	97
50–60	25	4	3
60–70	2	1	–
70–80	–	1	–
80–90	1	–	–
90–100	–	–	–
⟩100	–	1	2
Total	7 088	6 757	6 968
Average dose (mSv)	7·1	6·1	5·7

In summary, the parts of the fuel cycle that give the highest collective doses are reactors, fuel reprocessing and research and development, and the highest individual doses tend to be received by reprocessors and uranium miners.

The details of the average doses received by industrial radiographers, medical personnel and so on were given in the last chapter, and from a comparison with these we may conclude that some workers in the nuclear industry receive relatively high doses while others receive relatively low doses.

Finally, we must try to put the dose which the world's population can expect to receive from nuclear power into some perspective *vis-à-vis* the other sources to which we are exposed. Table 12.5 shows such an attempt made in 1977 by the Expert Committee of the United Nations. It gives the percentage of the total exposure attributable to a number of sources for the world's population.

Table 12.5. Relative global dose commitment from various radiation sources

Source of Exposure	Global dose commitment (% total exposure)
Exposure to natural sources	78
Commerical air travel	0·09
Production of phosphate fertilizers	0·009
Production of electricity by coal-fired power stations (10^6 MW(e))	0·004
Exposure to radiation-emitting consumer products	0·6
Production of nuclear power (8×10^4 MW(e))	0·1
Nuclear explosions (average 1951–1976)	6
Medical diagnostic radiation	15

Source: UNSCEAR Report, 1977 (modified).

The table shows that the major source of radiation is the natural environmental component and that all other sources are negligible except for medical radiology. In many ways the proportions given in this table are comparable with the data in table 11.5 for the UK population. All the values given in table 12.5 are crude averages and are subject to many qualifications that are outside the scope of this book. Just to give one example, the major component of the annual global collective dose from medical procedures is 5×10^5 man-gray and comes from the small number of the world's countries with developed radiological facilities; 2×10^4 man-gray coming from the majority of the world's population who live without the benefit of frequent radiological examination.

So the world's population is committed to a dose from the 1977/78 installed nuclear capacity (about 80 000 MW(e)) equal to 1·3 per cent of the natural background radiation. If the projected nuclear capacity for the year 2000—2×10^6 MW(e)—is attained the global dose commitment from nuclear power will rise to about 4 per cent of the natural background.

12.8. Reactor accidents

As we have just seen the radiation exposure of the population from the nuclear power programme is considerably less than that from such sources as medical radiology. Furthermore, during normal operations the radioactivity released by power stations cannot readily be distinguished from background levels. Why then are reactors usually sited in remote coastal regions? The answer is related in part to the possible consequences of a major accident involving the release of fission products, and in part dictated by the need for massive quantities of water to cool the spent steam from the turbines.

In this section we shall briefly examine the most difficult question: how safe are reactors?

From its inception the nuclear industry has been conscious of the hazards associated with radiation and stress has been placed on the prevention of unplanned occurrences. The possibility of accidents is minimized by systems of safety devices that give 'defence in depth' and so restrict the effect of any accident that does occur. However, absolute safety in so complex a system as a reactor is impossible, and significant reactor accidents have occurred in the USA, UK, Canada and Switzerland involving many types of reactor. But so far, in over 25 years of commercial nuclear plant operation, no member of the public has been injured as a result of reactor accidents, although the Windscale (Sellafield) fire in 1957 in the UK and the overheating and core and containment damage at the PWR (pressurized water reactor) at Three Mile Island, USA did pose acute public threats. Most of the unplanned release and accidents are trivial and as the Royal Commission on Environmental Pollution (1976) states "the administrative and practical arrangements that have been made for dealing with accidental release, both large and small, are good. . .".

Nevertheless, there remains the possibility of a major thermal reactor accident involving widespread contamination and some deaths among the public. It needs to be stressed that in a thermal reactor the fuel is near to its most active state and there is no way in which the nuclear assembly can go supercritical and explode like a nuclear weapon. The type of accident we are dealing with here involves the reactor safety features that control the cooling and running of the reactor. If they fail the temperature of the reactor core rises and the fuel melts, breaching its containment. Such a 'meltdown' results in the release of large quantities of radioactive materials. Two questions arise: how likely is such an event? And what are the likely consequences of such a major accident? We must tackle these questions.

The following example of a major credible accident at a 100 MW(e) nuclear power station is taken from the Report of the Royal Commission on Environmental Pollution (1976). It assumes that 10 per cent of the gaseous and volatile fission products are released from the reactor as a cloud of radioactive material. The reactor is sited in a semi-urban UK site. The main health hazard will be from ^{131}I (half-life 8 days), which will irradiate the thyroid, and ^{137}Cs (half-life 30 years), which will cause prolonged contamination of the countryside and buildings. Since the weather, especially the wind's direction and speed, will markedly affect the behaviour of the radioactive cloud, estimates of the number of people exposed and the hazards can only be expressed in terms of probability.

The inhalation of ^{131}I could cause thyroid cancer in people as far away as 24 km and there is a 20 per cent probability that over 10–20 years between 1 000 and 10 000 people could develop thyroid cancer. Only a fraction (<10 per cent) of these will be fatal cancers. The most probable outcome is 100–150 deaths from thyroid cancer plus a further 10–200 deaths from leukaemia and lung cancer over the same period. These figures could be 10 times higher or lower depending on the circumstances. The ^{131}I deposited on the ground would decay in a few weeks. However, the longer lived ^{137}Cs would be present in the soil and buildings for many decades. The radiation levels would necessitate evacuation for weeks, months or longer even for people up to 50 km from the reactor site. At distances less than 15 km the decontamination problem would be difficult and expensive. About 20 000 people might be severely inconvenienced by such an event. The assessment of the likelihood that such an event may occur is extremely difficult and involves a probabilistic method known as 'fault tree' or failure analysis technique. In the 'tree' all the possible sequences of malfunction are identified and all the consequences of each alternative malfunction are assigned probabilities. The cumulative probability of all failures are then assessed. Such an assessment was undertaken in 1975 by a committee of the US Nuclear Regulatory Commission and it took two years and cost two million dollars! The committee decided that the probability of a large accident involving the acute death of 1 000 people was very small—about 1 chance in 100 million years. The figure implies that a major reactor accident is far less likely than many other natural disasters (earthquakes, floods, volcanoes) or man-made disasters (dam failures, air crashes, chlorine releases). However, there are many uncertainties surrounding what such a statement even means since there has never been a major core meltdown in the past; and how can one ascribe a probability to such an event in the future?

As we have seen, a control failure in a thermal reactor can cause the meltdown of the fuel. The fuel sinks as a mass to the bottom of the reactor, but there is virtually no way in most thermal reactors in which this could lead to a supercritical situation. In contrast, if the plutonium fuel of a fast reactor is suddenly concentrated in a 'core collapse' meltdown it could possibly go supercritical producing a nuclear explosion, although the reaction would be very much slower than in a bomb. The explosion would vaporize the fuel releasing both volatile (iodine and caesium) and non-volatile (strontium, plutonium, etc.) components of the core. The consequences could be catastrophic, for example, if 10 per cent of the components of the core of a fast reactor were released the effects would be ten to a hundred times worse than those described above for a thermal reactor accident. Until the special problems of the fast breeder reactors are better understood it is unlikely that their widespread commercial use can be comtemplated.

Finally, in any type of reactor using a liquid coolant there is always the possibility of an explosion of the hot fuel mixes with the coolant. It is expected that such a non-nuclear explosion would be confined within the reactor containment, but it might not be, and it might so alter the geometry of a reactor core as to make it supercritical and cause a nuclear explosion.

12.9. Postscript: the nuclear debate

In this final section we shall brieflly touch upon some of the issues of the nuclear debate.

Man aspires to an ever-increasing quality of life and such aspirations require increasing energy supplies. This is the crux of the energy debate—a complex of economic, political, technical and ethical questions in which emotions often run high.

A country's economic progress, as measured by its gross national product, is closely associated with its per capita energy consumption. So, the question being asked in both developing and developed countries is how to bridge the increasing gap between energy demand and energy supply.

Fossil fuel supplies are finite and there will be a severe shortage of oil and gas (in about 20–30 years) and of coal (in about 200 years) if the world's increasing demand for energy continues. The increasing scarcity of fossil fuels will lead to steep price rises that will curb consumption and will force drastic changes in the patterns of consumption. Coupled with these economic factors there is the feeling that coal and oil should be used in the chemical industry and not simply burnt. So the inescapable question is: where is the world's energy to come from in the future? There is a wide measure of agreement throughout the developed world that nuclear fission is the only energy source at an adequately advanced stage to meet the huge gap between energy demand and supply that seems inevitable in 20–30 years time. It cannot of course solve the problem of finding a substitute for the liquid fuels used by most forms of transport.

At present, fission energy contributes about 1 per cent of the world's electricity supply. According to the forecasts of the United Nations, based on energy needs and population growth, there will have to be a 10-fold increase in the next 10 years in nuclear capacity and a 1 000-fold increase in the next 50 years. However, the world's known reserves of the fuel of current nuclear reactors, uranium, are only $4 \cdot 5 \times 10^6$ tonnes, so the projected programme of fission reactors will exhaust all economically available uranium in 2–3 decades. This leads logically to a future reliance on fast breeder reactors. This type of reactor is very efficient in its use of uranium but it also breeds new fuel, plutonium. Finally, there is the possibility of nuclear fusion: potentially the richest and cleanest source of nuclear energy, but it is much too early to know what part it will play in future energy strategy. There is unlikely to be a prototype fusion reactor until the next century.

The current consensus is that nuclear power is vital to our future energy needs although there are three options yet to be fully exploited. These are: (1) to reduce the level of energy consumption; (2) to adopt real conservation measures; and (3) to develop alternative sources.

Reduce energy consumption

Many people accept that there will inevitably be a slowing down in the demand for energy, but few are willing to accept a reduction in per capita consumption. In the most developed countries in the world per capita consumption is 20 times that of the developing world. Who are we to suggest that the world should make do with less energy?

Conserve energy

There is already a growing emphasis on the more efficient use of fuel and with spiralling costs conservation policies will be more widely adopted. But current energy conservation programmes cannot hope to bridge the energy gap forecast for the year 2000.

Alternatives to nuclear power

Among the possible sources are solar radiation, geothermal energy, wind and wave power. However, despite their ecological appeal relatively little effort or money has been put into these alternatives, and at present they remain economical in only a few places in the world. They all have their advocates and they all have great potential, if the technical problems can be solved.

The advocates of nuclear power believe it is essential to have a nuclear programme if we are to maintain our standard of living and if we are to avoid the wasteful use of valuable unreplaceable fossil fuels. They argue that strict observance of national and international radiological protection codes will minimize any potential hazards either to man or the environment. They would use much of the data in Chapter 11 and in this chapter to indicate the relative safety of the industry *vis-à-vis* other industries and the other hazards of the modern world.

However, from its inception the nuclear industry has had to face criticism, much of it ill-informed, but some of it cogent. We shall outline some of the arguments in these final paragraphs. There is a wide spectrum of opposition ranging from those who would prohibit it altogether to the more moderate critics who accept a limited programme as inevitable but are against a massive long-term commitment that will involve fast breeder reactors and the 'plutonium economy'. The opposition's fears centre around the scale of the future commitment with the dangerous implications of thousands of reactors, thousands of tonnes of plutonium and equally massive amounts of radioactive waste. Recent UN projections are for a 1000-fold increase in nuclear power capacity in the next 50 years. For example, in the UK in 2030 it is envisaged that there will be several hundred large fast breeders operating to produce the required 370 GW capacity! In the USA in 2020 there may be 2000 fast breeders and 60 fuel processing and fabrication plants. Such a commitment worries many people for many reasons and we can only touch on the issues of the nuclear debate, which include the following.

The radiological effects on the environment

Some critics argue that the radiological hazards posed by the release of radioactivity into the environment are not adequately understood and that the recommended limits set by the International Commission on Radiological Protection are not sufficiently conservative (see Chapter 11). The regulatory bodies would argue that the hazards from radiation are perhaps better understood than those of any other form of pollution. Most international scientific bodies would concur, but they would perhaps add that there is the need for continued vigilance and for research into areas where the recommended standards appear in any way suspect.

Reactor accidents

We have outlined the consequences of a serious accident in a thermal reactor and seen that the vaporization of the core contents of a fast reactor would do 10–100 times more damage. The designers claim that the probability of such an accident is between 10^{-5} and 10^{-6} per reactor per year and that the annual risk of death to an individual living close to a reactor is less than 1 in a million; compared with 1 chance in 8 000 per year of being killed in a road accident. The objectors would reply that individual mortality risks are perhaps less important than the overall effects of a serious nuclear accident. Furthermore, since power stations have been designed and are run by fallible human beings accidents do occur, and they will occur more frequently as the number of reactors increases. So, theoretical exercises that purport to show the improbability of a major reactor accident are irrelevant.

The authorities see these fears as irrational and feel that the safety record of the nuclear industry is second to none and entitles it to a degree of social acceptability *vis-à-vis* the other risks people live with in modern society.

The point is also made that reactors are ideal targets for sabotage in peacetime by terrorists and in wartime are strategic targets.

Radioactive waste

The critics query both the economics and the necessity of reprocessing the spent fuel elements. They are also worried by the problems associated with the need to isolate and contain high level wastes for thousands of years, which supposes the almost infinite stability of human societies throughout the world.

Similarly, no commercial reactor has ever been decommissioned and so the problem of what to do with the highly radioactive core, the containment structures and the associated machinery has still to be faced.

The authorities are confident that they can fix these problems so that future generations are not saddled with unacceptable levels of waste.

The problem of plutonium

The production of plutonium as a by-product of the nuclear energy industry raises the most immediate worries in the minds of many critics. Most nations will soon have nuclear power plants, and this will enable them to produce weapon-grade uranium and plutonium. This spread increases the likelihood of nuclear war. In the UK the civil nuclear programme was started primarily to provide plutonium and enriched uranium for weapons. It is feared that many countries are developing a nuclear war capacity under the guise of a civil nuclear programme. Such fears motivated the Israelis to bomb an Iraqi reactor in 1981.

Some proponents of nuclear power even deny the link between civil and military use of nuclear material, while others put their faith in the inspections and safeguards of the International Atomic Energy Agency and the Nuclear Non-Proliferation Treaty.

Despite such criticisms there is likely to be increasing reliance on nuclear power in the foreseeable future. Similarly, in the future there will be increasing interest in the exploitation of alternative energy sources.

Appendix
Table of radionuclides

Nuclide	*Symbol*	*Atomic number*	*Mass number*	*Half-life*	*Radiation* (MeV)
Actinium	Ac	89	227	27·7 yr	α(4·94); e⁻(0·043); γ
Americium	Am	95	241	433 yr	α(5·48); γ(0·06)
		95	242	16 h	e⁻(0·63); γ(0·04)
		95	243	7 370 yr	α(5·27); γ(0·75)
Argon	Ar	18	37	34 d	K†
Caesium	Cs	55	134	2·1 yr	e⁻(0·65, 0·09); γ(0·7)
		55	137	35 yr	e⁻(1·17, 0·518); γ(0·662)
Calcium	Ca	20	45	164 d	e⁻(0·256)
Californium	Cf	98	249	360 yr	α(6·3); γ(0·39)
Carbon	C	6	14	5 730 yr	e⁻(0·155)
Cobalt	Co	27	59		Stable
		27	60	5·3 yr	e⁻(1·48, 0·31); γ (1·332, 1·172)
Curium	Cm	96	242	163 d	α(6·11); γ(0·044)
		96	243	32 yr	α(5·78); γ(0·28); K
		96	244	18 yr	α(5·8); γ(0·43)
Deuterium	D	1	2		*see* Hydrogen
Hydrogen	H	1	1		Stable
		1	2		Stable
		1	3	12·3 yr	e⁻(0·018)
Iodine	I	53	125	60 d	K; γ(0·035)
		53	129	1·7×10⁷ yr	e⁻(0·013); γ(0·039)
		53	131	8·0 d	e⁻(0·6, 0·32); γs
Iron	Fe	26	59	46 d	e⁻(1·56, 0·46, 0·27); γs
Krypton	Kr	36	85	10·7 yr	e⁻(0·67); γ(0·52)
Manganese	Mn	25	54	303 d	K; γ(0·83)
		25	56	2·59 h	e⁻(0·7, 1·09, 2·88); γs
Neptunium	Np	93	239	2·33 d	e⁻(0·72); γs
Nickel	Ni	28	60		Stable
Niobium	Nb	41	94	2×10⁴ yr	e⁻(2·06); γ(0·87, 0·70)
Nitrogen	N	7	13	10·0 min	e⁺(1·19)

Nuclide	Symbol	Atomic number	Mass number	Half-life	Radiation (MeV)
Phosphorus	P	15	32	14·3 d	e⁻(1·718)
Plutonium	Pu	94	238	87 yr	α(5·59); γs
		94	239	2·44×10⁴ yr	α(5·24); γs
		94	240	6 600 yr	α(5·26); γs
		94	241	13·2 yr	α(5·15); γs
		94	242	3·8×10⁵ yr	α(4·98); γs
Potassium	K	19	40	1·28×10⁹ yr	e⁻(1·33); e⁺(1·50); γ(1·46); K
		19	42	12·4 yr	e⁻(3·55, 2·0); γ(1·52, 0·31)
Radium	Ra	88	223	11·4 d	α(5·97);
		88	224	3·6 d	α(5·80); γ(0·25)
		88	226	1 620 yr	α(4·78); γ(0·188)
Radon	Rn	86	222	3·82 d	α(5·486); γ(0·5)
Ruthenium	Ru	44	106	1·0 yr	e⁻(0·04)
Silver	Ag	47	105	40 d	K; γs
Sodium	Na	11	24	14·8 h	e⁻(1·39); γ(2·75, 1·38)
Strontium	Sr	38	90	28 yr	e⁻(0·55)
Sulphur	S	16	35	87 d	e⁻(0·17)
Thallium	Tl	81	204	3·80 yr	e⁻(0·76); K
Thorium	Th	90	232	1·4×10¹⁰ yr	α(4·08); γ(0·06)
		90	234	24·1 d	e⁻(0·19, 0·10); γs
Tritium	T	1	3		*see* Hydrogen
Uranium	U	92	235	7·1×10⁸ yr	α(4·68); γ(0·195)
		92	238	4·5×10⁹ yr	α(4·79); γ(0·048)
Xenon	Xe	54	135	9·3 h	e⁻(0·9); γ(0·25, 0·61)
Zinc	Zn	30	65	245 d	e⁺(0·33); K; γ(1·11)
Zirconium	Zr	40	95	64 d	e⁻(1·12); γ(0·72, 0·76, 0·25)

†K is a form of electromagnetic radiation. An orbital *K* electron is captured by the nucleus, thus raising the atomic number by 1. One of the outer shell electrons then fills the gap emitting the characteristic radiation. The radiation observed is consequently the complete X-ray spectrum of the daughter nucleus. This type of radiation emission is referred to as "*K* electron capture".

Glossary

Absorbed dose	Quantity of ionizing radiation. A measure of the energy imparted to a mass of material, e.g. tissue. The basic unit is the gray (Gy). 1 Gy = 1 joule per kilogram.
Activity	Quantity of a radionuclide which describes the rate at which decays occur in an amount of a radionuclide. The SI unit of radioactivity is the becquerel (Bq), which replaces the old unit, the curie (Ci). One becquerel corresponds to 1 disintegration of a radionuclide per second.
Becquerel	*See* Activity.
Cancer	Any malignant tumour. A tumour that tends to proliferate and spread indefinitely and to increase in virulence.
Cell cycle	The cycle of cellular events from one mitosis to the next. Four stages of mitosis—M, G_1, DNA synthesis (S) and G_2—are recognized as parts of the cycle.
Centromere	The constricted point on the chromosomes to which the spindle fibre is attached; the first part of the chromosome to move towards the pole at mitosis or meiosis.
Chromosome	Filamentous structure in the cell nucleus along which genes are located.
Collective dose	Usually refers to 'collective effective dose equivalent' obtained by multiplying the average effective dose equivalent by the number of persons exposed to a given dose. Expressed in man-sievert (man-Sv).

231

Curie (Ci)

See Activity. A radionuclide was said to have an activity of 1 curie (Ci) if it transformed at a rate of $3 \cdot 7 \times 10^{10}$ disintegrations per second.

D_0

The dose required to reduce survival by a factor of e^{-1} (0·37) along the exponential part of the survival curve.

Deoxyribonucleic acid (DNA)

The nucleic acid found especially in the cell nucleus; the genetic material.

Direct radiation effects

Damage caused by the absorption of radiation energy directly in a critical biological site or target (*see* figure 1.1).

Dose

General term for the quantity of radiation. *See* Absorbed dose, Dose equivalent, Effective dose equivalent, Collective dose.

Dose equivalent

Quantity obtained by multiplying the absorbed dose (in gray) by a quality factor to allow for the degree of effectiveness of particular types of radiation. The unit is the sievert and the quality factor (*see* table 1.6) for X- or γ rays or β particles is 1, for neutrons 10 and for α particles 20.

Effective dose equivalent

Quantity obtained by multiplying the dose equivalents to a tissue by the appropriate risk weighting factor (*see* table 11.6) for that tissue. Expressed in sievert.

Electron volt (eV)

The amount of energy gained by a particle of charge e ($-1 \cdot 6 \times 10^{-19}$ C) when accelerated by a potential difference of one volt. $1 \text{ eV} \approx 1 \cdot 6 \times 10^{-19}$ J.

Exponential survival curve

A survival curve without a threshold or shoulder region and which is a straight line on a semi-logarithmic plot (*see* figure 3.1*b*).

Fission products

Nuclides or radionuclides produced as a result of nuclear fission; the process in which a nucleus splits into two or more nuclei with the release of energy.

Free radical

A group of atoms that is normally in combination with other atoms, but which can exist independently and as such is chemically very reactive.

Fusion

Thermonuclear fusion occurs when two or more light nuclei coalesce to form a heavier nucleus with the release of energy.

Gamma ray	A discrete quantity of energy emitted by a radionuclide, lacking mass or charge, that is propagated as a wave.
Gene	The unit of inheritance; usually a portion of a DNA molecule that codes for some product such a protein.
Genetic effects of radiation	Damage done to the cells of the reproductive system, which may be heritable.
Genetically significant dose	The dose that, if given to every member of a population, should produce the same hereditary harm as the actual doses received by the individuals. Expressed in sievert this dose takes into account the child-bearing potential of those receiving the dose.
Genome	The haploid set of chromosomes with their associated genes.
Gonads	The ovaries or testes.
Gray	*See* Absorbed dose.
Half-life	The time taken for the activity of a radionuclide to decay to half its initial value. Symbol $t_{\frac{1}{2}}$.
Haploid	Having only one of each type of chromosome; as is usually the case in gametes (oocytes and spermatozoa).
Indirect radiation effects	Damage cause by the absorption of radiation energy in the vicinity of a critical site or target (*see* figure 1.11).
Ion	An electrically charged atom or group of atoms.
Ionization	The process whereby a neutral atom or molecule acquires an electric charge.
Isotope	Nuclides with the same number of protons but different numbers of neutrons.
Linear energy transfer (LET)	The rate of energy loss along the track of an ionizing particle, usually expressed in keV per micrometre of track (keV μm^{-1}).
Mitosis	Process of nuclear division in which chromosomes move along a spindle resulting in two new nuclei with the same number of chromosomes as the original nucleus.

Mitotic index	The proportion of cells in a population in mitosis at any given time.
Mutation	A relatively stable change in the DNA of the cell nucleus. Mutations in the germ cells of the body (ova and sperm) may lead to inherited effects in the offspring. Mutations in the somatic cells of the body may lead to effects in the individual, e.g., cancer.
Nuclear reactor	A structure in which nuclear fission may be sustained in a self-supporting chain reaction. In thermal reactors the fission is produced by thermal neutrons, in fast reactors by fast neutrons.
Nucleic acid	A class of organic acids that play a role in protein synthesis, the transmission of hereditary traits and the control of cellular activities.
Nucleus	The membrane-bound organelle in a cell that contains the chromosomes.
Nuclide	A species of atom defined by the number of its protons and neutrons.
Oxygen enhancement ratio (OER)	The ratio of the radiation dose given under anoxic conditions to produce a given effect relative to the radiation dose given under fully oxygenated conditions to produce the same effect (*see* figures 8.5, 8.7, 8.8).
Ploidy	Relates to the number of sets of chromosomes in a cell. *Diploid* cells have two sets of chromosomes, a chromosome complement twice that found in the gametes. *Tetraploid* cells have four sets of chromosomes.
Pressurized water reactor (PWR)	A thermal nuclear reactor which uses water both as a moderator and coolant.
Purines	Organic bases with carbon and nitrogen atoms in two interlocking rings; components of nucleic acids and other biologically active substances (*see* figure 2.2).
Pyrimidines	Nitrogenous bases composed of a single ring of carbon and nitrogen atoms; components of nucleic acids (*see* figure 2.2).
Rad	The old unit of absorbed dose, equivalent to an energy absorption of 10^{-2} J kg^{-1}. Superseded by the gray (*see* Absorbed dose).

Radioactive waste	Unwanted radioactive materials in any form. Often categorized in the nuclear power industry into low level, intermediate level and high level waste.
Radioactivity	The property a radionuclide possesses of spontaneously emitting ionizing radiation. Materials containing radionuclides are said to be radioactive.
Radiotherapy	The treatment of disease with ionizing radiation.
Rem	The old unit of dose equivalence, superseded by the sievert (*see* Effective dose equivalent).
Relative biological effectiveness (RBE)	The ratio of the absorbed doses of two radiations required to produce the same biological effect (*see* figure 8.3).
Reproductive death	The loss of the proliferative ability of a cell. Commonly restricted to those cells having an indefinite capacity to divide (*see* Chapter 3).
Sensitizing agent	A substance that increases the biological effectiveness of a given dose of radiation.
Sex-linked	Of genes: located on the X or Y chromosome.
Sievert	*See* Effective dose equivalent.
Somatic	Pertaining to the body; to all cells except the germ cells.
Somatic effects of radiation	Damage that is apparent during the lifetime of the organism, exclusive of effects on the reproductive system.
Tumour	A swelling or enlargement; especially one due to the pathological proliferation of cells and marked by an independence of the normal mechanisms that control growth, and having no physiological function.
X chromosome	The female sex chromosome.
X-ray	Discrete quantity of energy lacking mass or charge; propagated as a wave and produced in an X-ray machine.
Y chromosome	The male sex chromosome.

Further Reading

Chapter 1.
Johns, H.E. and Cunningham, J.R., 1977, *The Physics of Radiology*, 3rd edition (Springfield, Ill., USA: Charles C. Thomas).
Meredith, W.J. and Massey, J.B., 1974, *Fundamental Physics of Radiology* (Bristol, UK: J. Wright and Sons).

Chapters 2, 3, 4 and 5.
Elkind, M.M. and Whitmore, G.F., 1967, *Radiobiology of Cultured Mammalian Cells* (New York: Gordon & Breach).
Chadwick, K.H. and Leenhouts, H.P., 1981, *Molecular Theory of Radiation Biology* (New York: Springer-Verlag).
Alper, T., 1979, *Cellular Radiobiology* (Cambridge: University Press).

Chapter 6.
Bond, V.P., Fliedner, T.M. and Archambau, J.O., 1965, *Mammalian Radiation Lethality* (New York: Academic Press).
United Nations Scientific Committee on the Effects of Atomic Radiation (UNSCEAR), 1977, *Sources and Effects of Ionising Radiation*, UN Publication E.77.ix.1.

Chapter 7.
UNSCEAR Report, 1977 (ibid.).

Chapters 9 and 10.
UNSCEAR Report, 1977 (ibid.).
Meyn, R.E. and Withers, H.R. (eds.), 1980, *Radiation Biology in Cancer Research* (New York: Raven Press).
Duncan, W. and Nias, H.W., 1977, *Clinical Radiobiology* (Edinburgh: Churchill Livingstone).
Hall, E.J., 1978, *Radiobiology for the Radiotherapist* (New York: Harper and Row).

Chapter 11.
International Commission on Radiation Protection (ICRP), 1977, *Recommendations of the International Commission on Radiation Protection,* ICRP Publications 26 and 27 (Oxford: Pergamon Press).
National Radiological Protection Board of the United Kingdom, 1981, *Living with Radiation* (Harwell, UK: NRPB).

Chapter 12.
Pentreath, R.J., 1980, *Nuclear Power, Man and the Environment,* Wykeham Science Series No. 51 (London: Taylor & Francis).
Royal Commission on Environmental Pollution, 1976, *Sixth Report: Nuclear Power and the Environment* (London: H.M.S.O.).

Index

239